Peter Crut £1

P

The Essential Writings of Erasmus Darwin

The Essential Writings of

ERASMUS DARWIN

Chosen and edited
with a linking commentary, by

Desmond King-Hele

MacGibbon & Kee

First published 1968 by MacGibbon & Kee Ltd
3 Upper James Street Golden Square London W1

Printed in Great Britain by Ebenezer Baylis & Son Ltd
The Trinity Press, Worcester, and London

Contents

Illustrations

The
Essential Writings of
Erasmus Darwin

Introduction

DR ERASMUS DARWIN, who lived from 1731 until 1802, was famous in his own day for a wide variety of achievements. Recognized as the finest physician in England, he had a large practice centred on Lichfield and Derby, and a reputation for giving poor patients food and blankets instead of charging fees. In the 1790s he won a dazzling but transient fame as a poet, with his *Botanic Garden*, which strongly influenced Wordsworth, Coleridge and Shelley. He was noted for his many inventions, ranging from a speaking-machine and a horizontal windmill to novel designs for carriages, drill ploughs and canal lifts. He was extremely sociable: not only did he enjoy a full and happy family life—he married twice and had fourteen children—but he was also a founder of several societies and a close friend and adviser of the Midland industrialists who did so much to bring about the Industrial Revolution, men like Boulton, Keir, Watt and Wedgwood. Darwin was also regarded as the most knowledgeable scientist of his time, the author of massive treatises on animal and plant life, *Zoonomia* and *Phytologia*.

Today Darwin is perhaps best known for other achievements—his original ideas and discoveries in several branches of science. He foreshadowed in some detail the theory of evolution later developed by his grandson Charles, and in his long poem *The Temple of Nature* traced the progress of life from microscopic specks in primeval seas to its present culmination in man. Darwin was the first to explain the principles of artesian wells and of fertilizers; he discovered how clouds form, by the adiabatic expansion of moist rising air; he gave an almost complete description of photosynthesis, and was a pioneer in the humane treatment of mental illness, to quote but a few of his other varied achievements.

Darwin belongs with Leonardo da Vinci and Goethe, in the small band of great all-rounders who excelled in both practical and theoretical work, in both literature and science. No once since Darwin's day has equalled him in the scope and variety of his accomplishments, and in our fragmented culture few can even appreciate the full range of his achievements.

In this book I have tried to sample his achievements widely, by quotation from his own writings. Those writings deserve to be brought into circulation again, for there was more than a little truth in the *Monthly Magazine*'s comment in 1802, that no man 'had a more imperial command of words'. Darwin's skill in cramming scientific facts into rhyming couplets is unrivalled, though sometimes the result is scarcely poetic. But he expressed himself even better in prose: some of his letters are as lively and pungent as any in the language, and this style often animates his scientific writings.

I have tried to choose the most interesting passages and have added a linking commentary to summarize omitted sections. My aim has been to make the text as continuous and readable as possible, while providing an anthology which gives a fair impression of Darwin's writings and is a balanced mixture of science, poetry, biography and social history.

1 Life

A humanist, physician to all men.
KEATS; *Fall of Hyperion*

ERASMUS DARWIN was born at Elston Hall, about ten miles north-east of Nottingham, on 12 December 1731. He was the fourth son of Robert Darwin, an 'honest and industrious' lawyer with a taste for archaeology. Although he had inherited Elston Hall, Robert was never wealthy. He had married Erasmus's mother, Elizabeth Hill, as she then was, in 1724. She is said to have been a capable and scholarly woman, strong in mind and body: she lived to be ninety-five, and 'to the last day of her life got up to feed her pigeons'.

Erasmus was sent to Chesterfield School at the age of nine and seems to have been quite happy there. In 1750 he went up to St John's College, Cambridge, as an Exeter scholar; but he had to live very frugally because his father was already supporting two other sons at Cambridge. Erasmus passed his examinations in classics, mathematics and medicine, and took his B.A. degree in 1754. He also wrote poems, two of which were published: one was entitled *The Death of Prince Frederick*; the other was in praise of Gurney's phonetic shorthand, which he learnt in order to take lecture notes—the beginning of a life-long interest in phonetics.

To complete his medical training he went for two years to Edinburgh, then at the height of its renown as a medical school. There he became friendly with James Keir, later famous as a chemist and a leading member of the Lunar Society of Birmingham.

Darwin completed his studies at Edinburgh in 1756, and in the autumn, with all the erroneous eighteenth-century medical know-how at his fingertips, he was ready to be let loose on the public. It was at Nottingham that his splendid career began, or rather failed to begin, because no patients came to him in his native county town. At this critical moment Darwin showed his mettle by taking up the case of a poor local patient who had been overcharged by a London surgeon. Eventually the surgeon paid back two-thirds of the six-guinea fee. The professional men of Nottingham, like the Sheriff of old, must have been wondering who would be exposed next by this new Robin Hood. But in November 1756 he turned his back on Sherwood Forest and left for Lichfield, which was to be his home for twenty-five years.

Lichfield, then better known than its growing industrial neighbour Birmingham, fifteen miles to the south, is an ancient cathedral city: the list of Bishops dates from 656 and the first church on the cathedral site was consecrated in 700. Thirteen centuries have brought many changes, but most of them had already occurred by 1756, and the Cathedral Close, with Darwin's house on its outskirts, still remains much the same as in the eighteenth century. Darwin came to Lichfield with a letter of introduction to Canon Seward, who combined his duties at the Cathedral with the editing of Beaumont and Fletcher's plays.

Darwin's medical career at Lichfield began brilliantly. A young man from a leading county family, Mr Inge, lay dangerously ill and other doctors had written him off as a hopeless case: Darwin was called in as a last resort, and cured him. Mr Inge lived on to enjoy a useful career of public service—and become ancestor of the twentieth-century Dean of St Paul's. After this flying start, Darwin never looked back. His practice grew steadily, while his despairing rivals complained that his methods were rash and experimental.

A few months after arriving in Lichfield Darwin fell in love with Mary Howard, the eighteen-year-old daughter of a cathedral precentor. They were married in December 1757, and in 1758 went to live in the house on Beacon Street at the western edge of the Cathedral Close which was Darwin's home for over twenty years. (Fig. 5 shows the house as it is today.) Mary was described as 'blooming and lovely', with a mind of 'native strength' and an 'awakened taste for the works of imagination'. Their marriage was a very happy one, marred only by Mary's continual ill-health.

The Cathedral Close in the 1760s was not as dull as its name suggests. Canon Seward's lively daughter Anna was in her early twenties, and with Darwin's encouragement developed her talent for writing verse: she became known as 'The Swan of Lichfield'. Many years later she wrote down her memories of Darwin in a book that is generous yet also malicious, and quite inimitable in its stilted phraseology.

What was Darwin like in his younger years at Lichfield? One answer is provided by the portrait reproduced here as Fig. 1. Anna tells us 'he was somewhat above the middle size, his form athletic, and inclined to corpulence; his limbs too heavy for exact proportion'.[1]* Edgeworth declares, more bluntly: 'He was a large man, fat and rather clumsy.'[2] He was cheerful and enjoyed very good health—as he needed to if he was to succeed as a country doctor. He stammered quite badly, but was nevertheless a domineering talker in society, noted for his wit and occasionally, if he was thwarted, for his sarcasm. When Dr Johnson returned to Lichfield, his home town, he was none too pleased to find a sage as weighty and positive as himself in possession of the field, especially since Darwin was unconventional enough to despise both Christianity and alcoholic drink. In private conversation Darwin was not at all domineering, and so knowledgeable that his advice was eagerly sought by the many famous men who were among his friends.

Darwin was at his most energetic in the 1760s, when he built up the circle of friends which developed into the Lunar Society. He did this, we should remember, while spending much time, at all hours of the day and night, being 'jostled and joggled' in his carriage as it bounced over the vile roads *en route* to his patients. And if the weather was really bad, the carriage got stuck in the mud and he had to travel on horseback, arriving soaked to treat a patient who quite possibly had a contagious disease with no known cure. Even to survive was quite an achievement.

Darwin's earliest close friend in the Birmingham region was Matthew Boulton[3] (1728–1809), then a struggling manufacturer of buckles. Darwin and Boulton probably met in 1758, and Benjamin Franklin's visit to Birmingham in that year stimulated their interest in science. Boulton was keen to expand business in the new factory he was building at Soho, on the northern outskirts of Birmingham, and in about 1763

* Numbered references are listed at the end of the book, p. 201. They are intended primarily to provide material for further reading.

Darwin suggested making a steam-carriage, with twin cylinders and rear-wheel drive. Darwin was madly keen on the idea, but Boulton was not convinced it was practical. However, Darwin's enthusiasm helped to interest Boulton in steam engines and paved the way for his famous partnership with Watt. Even if he had never met Watt, Boulton would probably have earned his titles of 'The Father of Birmingham', and 'the first manufacturer of England', for he also won renown for his superb work as a silversmith and for his coining-machines, which dismayed the counterfeiters and made money for half the world.

Darwin's 'favourite friend' in the late 1760s was Dr William Small (1734–1775), who arrived from America in 1765 with a letter of recommendation from Franklin. For the previous seven years Small had been Professor of Natural Philosophy at the College of William and Mary in Virginia, and Thomas Jefferson, one of his pupils, declared that Small's teaching 'probably fixed the destinies of my life'. In the next ten years Small earned the highest praise from Darwin and all his friends for his profound learning and personal charm, but he remained entirely behind the scenes and left no record of his powers. He suffered continual ill-health, and died in 1775 at the age of forty-one.

In 1765 Josiah Wedgwood[4] was thirty-five years old and had been running a pottery works of his own for six years. The export trade he was trying to build up was seriously handicapped by the high cost of road transport, and the numerous breakages. Being immensely energetic, Wedgwood decided to promote the first major English canal, the Trent-and-Mersey, and in 1765 he found a staunch ally in Darwin, who helped by writing pamphlets and enlisting support from influential patients. A Parliamentary Bill for the canal was passed in 1766. Darwin also took a keen interest in Wedgwood's pottery works and designed a horizontal windmill for his new factory at Etruria. Wedgwood was greatly indebted to Darwin and his circle of friends: without them and their expert advice in chemistry and mineralogy, he would not have achieved his unique success in simultaneously advancing science, art and industry.

The next addition to Darwin's circle of friends, Richard Lovell Edgeworth[5] (1744–1817), seemed to be heading for the way of life appropriate to a young Anglo-Irish landed gentleman, when, in 1765, he heard that Darwin had designed a carriage peculiarly resistant to overturning. Edgeworth thereupon decided to devote himself to invention, and in 1766 he called on Darwin, who was at once impressed both by his charm and by his inventiveness. Their friendship lasted to the end of Darwin's life. Edgeworth was as ingenious and prolific an inventor as Darwin, and he should have received credit for the telegraph and 'Macadamized' roads, as well as for his scientific work on the theory of carriage springs and the air resistance of projectiles. Inventions were not the only subjects of their discussions: they were also both keenly interested in education, and the influential volumes of Edgeworth's *Practical Education* (1798) more than matched Darwin's little book on *Female Education* (1797).

This interest in education was stimulated by Darwin's meeting in 1766 with Rousseau, who was then living nearby at Wootton Hall. Apparently they wrote to each other for several years, but the letters have not been preserved. Rousseau's *Emile* particularly interested Darwin's friends: Wedgwood took some notice of Rousseau's ideas in educating his children, and Edgeworth brought up his eldest son on Rousseauistic lines, not with very happy results.

It was in 1765 that James Watt[6] (1736–1819) made his discovery of the separate condenser for improving the steam engine. When he came to England in 1767 with the invention still undeveloped, he visited Darwin, the steam-carriage enthusiast, and confided the secret of his invention. At this time Watt's engine was being unsuccessfully developed by Dr John Roebuck. In 1773 Roebuck became bankrupt, and Boulton, strongly urged on by Small and Darwin, took over Watt's patent: thus began the epoch-making partnership of Boulton-and-Watt, whose steam engines were prime movers of the Industrial Revolution. From their very first meeting Darwin and Watt became firm friends, and Watt relied on Darwin for encouragement, ideas and advice.

In the same year that Watt visited England, 1767, Darwin's old friend from his Edinburgh days, James Keir[7] (1735–1820), decided to retire from the Army, and came to live at West Bromwich. After a few years as manager of a glass factory he set up the Tipton alkali works, where he succeeded in making caustic soda from salt on a commercial scale many years before the Leblanc process was heard of. Keir used a clever method relying on dilute solutions: the details have only recently been elucidated by J. L. Moilliet.[8] The Tipton works may be said to mark the birth of the modern chemical industry.

In the late 1760s and early 1770s this group of friends—Boulton, Darwin, Small, Wedgwood, Edgeworth, Watt, Keir, and afterwards others—would occasionally meet at one or other of their houses. Gradually the meetings became more regular and whenever possible were held monthly a little before the time of the Full Moon, so that the visitors could have some light to help them on their way home in the evening.

These 'Birmingham Philosophers' who met at the Full Moon became known as the Lunar Society of Birmingham.[9] But the Society had no formal organization and only flourished because the members were so active and reacted upon each other in such a stimulating way. The Society had grown out of the friendship of Boulton and Darwin, and if we formalize the informal by calling them President and Vice-President, Small was the Secretary: Darwin introduced most of the new members, but it was Small who kept them in the Lunar web. The Lunar Society was at the very root of the Industrial Revolution in England and hence of the technological modern world. It might well be called the greatest scientific non-organization of all time.[10] In 1966 the Society's bicentenary was marked by exhibitions and a medallion, shown in Fig. 3.

In 1768 Darwin suffered a serious accident, when he was pitched out of his carriage and injured his right knee so badly that he was ever after slightly lame.

Apart from this miscarriage, the 1760s were almost free of troubles: Darwin was fully and happily occupied with his friends, his patients and, most of all, with his family. There were five children of his first marriage, and three survived the perils of an eighteenth-century infancy, Charles (born 1758), Erasmus (1759), and Robert (1766). Unfortunately their mother's health was never strong, and 'the frequency of her maternal situation . . . had probably a baneful effect', as Anna Seward put it. Mary Darwin's disease is not known, but Dr T. K. With has suggested that it may have been acute intermittent porphyria. Whatever it was, the disease seems to have been hereditary, for somewhat similar symptoms appeared in several of her descendants, including her grandson Charles Darwin.

In 1770, when she was only thirty-one, Mary Darwin died. Anna Seward often sat with her in her last weeks, and recorded Mary's generous tributes to her husband. His 'indulgence to all my wishes, his active desire to see me amused and happy, proved incessant. . . . Married to any other man, I do not suppose I could have lived a third part of those years which I have passed with Dr Darwin; he has prolonged my days, and he has blessed them'.[11]

During the 1770s Darwin kept up his medical practice and his friendships with the Lunar group as vigorously as ever, and his energy overflowed into various new channels. In 1771 he was hard at work on his speaking-machine. In 1772 he had long discussions with Brindley and Wedgwood about joining the Grand Trunk Canal to Lichfield with a narrow-gauge canal, but in the end he abandoned the idea. With his friend Brooke Boothby he founded the Lichfield Botanic Society, which later produced several voluminous translations of the works of Linnaeus: a great deal of work was needed for these books and it is remarkable that the Society consisted of only three members—Darwin, Boothby (who was a mere dilettante and did nothing), and the Rev. Joseph Jackson, a proctor at the cathedral. Darwin's botanic interests also flowered in a piece of land a mile west of Lichfield, where in about 1778 he established a botanic garden, famous locally at the time, and later nationally, as the inspiration of his poem of the same name.

During the 1770s there were a few changes in the Cathedral Close. Darwin's sister Susannah came to act as housekeeper, while Anna Seward flirted with him and they exchanged amorous letters written in the names of their cats. Darwin had other female company too: 'his affection for Venus was retained to the last period of his life', as a discreet early biographer remarked. Darwin had two illegitimate daughters, known as the Misses Parker, in the 1770s. They were brought up in his house and later established a very successful boarding school at Ashbourne on the edge of Sir Brooke Boothby's estate.

A new arrival at Lichfield in 1770, and another recruit for the Lunar Society, was Thomas Day[12] (1748–1789), an unkempt and eccentric friend of Edgeworth, who soon became a friend of Darwin too. Day was completely unconventional and yet wrote that most proper of books, *Sandford and Merton.* He put Rousseau's educational ideas to a severe test by adopting two girls from an orphanage and training them, in the hope that one would make a good wife. In 1770 Day was hard at work teaching the more amenable of the two girls, Sabrina. In this task of unique difficulty, who better to give advice than Darwin? So Day came to Lichfield, and remained there for a time even after he had realized that Sabrina would not suit him. However it was the ever-helpful Dr Small, and not Darwin, who eventually found a wife for Day.

When Dr Small died in 1775, Darwin invited Dr William Withering[13] (1741–1799) to take over Small's medical practice. Withering also became a useful member of the Lunar Society, being especially knowledgeable in chemistry and botany.

In 1780, when the Lunar Society was at its strongest, it received an important recruit, Dr Joseph Priestley[14] (1733–1804), perhaps the greatest experimental scientist of the day. Darwin, who was already acquainted with Priestley, helped to organize a subscription which enabled Priestley to pursue his scientific work. This freedom from financial stress, together with the encouragement and companionship of the Lunaticks, made Priestley's years at Birmingham the happiest of his life.

The year 1778 brought a personal tragedy for Darwin. His eldest son Charles, who showed brilliant promise as a medical student at Edinburgh, cut his finger while dissecting: the wound turned septic and he died in May, aged nineteen. Darwin regarded this as the cruellest blow of his life. In memory of his son he edited and published Charles's medical dissertation, which had won him the first gold medal of the Aesculapian Society.

The year 1778 also marked another and quite different crisis in Darwin's life. He fell violently in love with Mrs Elizabeth Pole (Fig. 2), the young and beautiful wife of Colonel Sacheverel Pole, of Radburn Hall, near Derby, head of one of the leading families in Derbyshire and a hero of the Battle of Minden.

Elizabeth, then 'in the full bloom of her youth and beauty' according to Anna Seward, was a witty, carefree and unconventional woman, the illegitimate daughter of the second Earl of Portmore. In looks as well as temperament, she resembled her grandmother Catherine Sedley, who was the mistress of King James II before marrying the first Earl of Portmore.

Since Darwin was fat, ugly, lame and nearly fifty, his attentions to Mrs Pole seemed likely to be fruitless—the unpresentable in full pursuit of the unattainable. But Darwin had other ideas, and, as usual, he succeeded. He began his wooing of Elizabeth Pole by sending her a tea-vase, with appropriate couplets. This is the beginning of an astonishing courtship in verse, which Anna Seward compared with Petrarch's verses to Laura. Though Darwin's poems do not match Petrarch's, their outcome was more favourable: Colonel Pole died in November 1780, and Darwin, brushing aside competition from many of the handsome young bachelors of Derbyshire, married Elizabeth in 1781.

After his marriage Darwin moved to Radburn Hall, since Elizabeth objected to Lichfield, or possibly just to Anna Seward. The Darwins remained for two years at Radburn, which is about four miles west of Derby, and then in 1783 moved into a house near the river, in Full Street, Derby. Darwin seems to have had no regrets at leaving his friends and botanic garden in Lichfield, and partly cutting himself off from the Lunar Society. His second marriage was as happy as the first, and the seven children of the marriage were fortunate, as they readily admitted, in being able to grow up in such happy surroundings.

In his years at Derby Darwin attained legendary fame as a doctor. Patients came to him from far and wide, and the stories about his almost magical powers became part of Midland folk-lore. King George III wanted Darwin to be his physician, and would no doubt have been better treated during his spells of mental derangement if he had been under Darwin's care: but Darwin would not move south.

One of the simplest yet most incredible stories about him concerns alcohol. Darwin strongly opposed alcoholic drink, because of its effects on health. And apparently he persuaded the hard-drinking squires of Derbyshire to heed him. Maria Edgeworth reported that most of the gentry in Derbyshire and neighbouring counties had become water-drinkers, and Anna Seward is even more specific: 'It is well known that Dr Darwin's influence and example have sobered the county of Derby', she declares, 'intemperance in fermented fluid of every species is almost unknown amongst its gentlemen'.[15] Could there be a stronger proof of his prestige (even if they did keep secret bottles locked away)?

Soon after he moved into the town in 1783, Darwin founded the Philosophical Society of Derby, on rather similar lines to the Lunar Society, except that its other members were lightweights by comparison with the Lunaticks. However, the Society built up a useful library and sponsored programmes of lectures until it was merged with the Derby Museum and Library in 1857.

After 1780 the Darwin and Wedgwood families grew even more friendly. Darwin keenly encouraged Wedgwood in his copying of the Portland Vase, his finest work. Darwin was sent the first good copy and produced what is still considered the best explanation for the scenes depicted on the Vase. In 1796 Darwin's son Robert, who had been in medical practice at Shrewsbury for ten years, married Wedgwood's daughter Susannah: Charles Darwin was one of their sons.

In the 1780s Darwin also became friendly with Samuel Galton (1753–1832), a Birmingham manufacturer who was interested in science and a member of the Lunar Society. Darwin's eldest daughter by his second marriage, Violetta, married Galton's son Samuel Tertius. One of their sons was Sir Francis Galton, and their daughter Lucy linked the Lunar families even more widely by marrying Keir's grandson James Moilliet.

The years from 1784 to 1799 passed contentedly for the Darwins, and only one of their seven children died in infancy. But in 1799 Darwin's elder surviving son, Erasmus, drowned himself in the Derwent one stormy night. Erasmus junior practised as a solicitor, but he was more interested in collecting coins and in statistics; and he shared his father's liking for chess, as Fig. 4 shows. Although he had been rather eccentric for a few months beforehand, his death at the age of forty was quite unexpected, and saddened his father's last two years.

Darwin moved house for the last time in 1802, to Breadsall Priory four miles north-east of Derby, a pleasant country mansion at the head of a long valley with a 'talkative stream' running down it. Here Darwin died of a heart attack on 18 April 1802, aged seventy. He was buried in Breadsall Church. Elizabeth lived on for nearly thirty years at the Priory, which became known as 'Happiness Hall' to her Galton grandchildren.

In the 1790s Darwin had continued to meet his old friends Boulton, Keir, Wedgwood, Watt and Edgeworth, and he also made a few new ones, notably Dr Thomas Beddoes[16] (1760–1808), an extreme radical who wrote pamphlets against the Church. In 1798, with financial help from Tom Wedgwood and Edgeworth, Beddoes founded the Pneumatic Institution at Clifton, where he hoped to discover a cure for consumption and instead discovered Humphry Davy. Some of the letters between Darwin and Beddoes were published, and this association with Beddoes, together with the theories propounded in *Zoonomia*, gave Darwin the reputation of being an atheist or at the very least a materialist.

One of the younger generation who failed to become friendly with Darwin was Coleridge, who visited him in 1796. Coleridge disliked Darwin's evolutionary ideas, his verse, and his scorn of Christianity; when they met, Darwin bantered Coleridge on the subject of Unitarianism, which Darwin said was 'a feather-bed to catch a falling Christian'. Nevertheless, Coleridge, himself famous for his wide and curious knowledge, wrote that 'Dr Darwin possesses, perhaps, a greater range of knowledge than any other man in Europe, and is the most inventive of philosophical men'.[17]

In succeeding chapters we shall see something of that inventiveness, as displayed in his various writings. Chapter 2 gathers excerpts from his many lively letters, and Chapter 3 records opinions and stories about him by his contemporaries. Chapter 4 covers his influential little book on girls' education. In Chapter 5 we dip into the massive medical volumes of *Zoonomia*. Chapter 6 gives his views on evolution, which was for him not so much a theory, more a philosophy of life. Chapter 7 is full of plant life—excerpts from *Phytologia*, perhaps the best of his books. Chapter 8 is devoted to his mechanical inventions, with descriptions and sketches taken largely from manuscript sources. Chapters 9 and 10 provide an anthology from his long and encyclopaedic poems, *The Botanic Garden* and *The Temple of Nature*; and Chapter 11 shows their influence on the Romantic poets. Chapter 12 records his discoveries and speculations in physical science, from the notes to his poems and his papers in the *Philosophical Transactions* of the Royal Society.

2 Letters

The first *literary* character in Europe, and the most original-minded Man. COLERIDGE on Darwin

DARWIN is one of the liveliest of letter-writers, and quite free from the formality which lays its numbing hand on so many eighteenth-century letters.

This liveliness pervades even his first recorded letter, dating from 1748, when he was sixteen. His sister Susannah had written describing her abstinence in the first days of Lent, and telling him she intended to eat pork because a 'learned Divine' said that hogs had been fish ever since the Devil entered into them and they ran into the sea. Erasmus replied as follows:*

Dear Sister,

I receiv'd yours about a fortnight after ye date yet I must begg to be excused for not answering it sooner: besides I have some substantial Reasons, as having a mind to see Lent almost expired, before I would vouch for my Abstinence throughout ye whole; and not having had a convenient oppertunity to consult a Synod of my learned friends about your ingenious Conscience, and I must inform you we unanimously agree in ye Opinion of ye Learned Divine you mention, that Swine may indeed be fish, but then they are a devillish sort of fish; and we can prove from ye same Authority that all fish is flesh whence we affirm Porck not only to be flesh but a devillish Sort of flesh; and I would advise you for Conscience sake altogether to abstain from tasting it; as I can assure You I have done, tho' roast Pork has come to Table several Times; and for my own part I have lived upon Puding, milk, and vegetables all this Lent; but don't mistake me, I don't mean I have not touch'd roast beef, mutton, veal, goose, fowl, etc. for what are all these? All flesh is grass! . . .

* In quotations from Darwin's letters I have retained the original spellings whenever practicable.

(To Susannah Darwin, March 1748. C. Darwin, *Life of Erasmus Darwin*, pp. 9–10*)

Obviously Darwin developed at an early age his likings for food and for banter about religion.

Four days before his marriage to Mary Howard in 1757, Darwin wrote her a long letter from Darlaston. He is keen about Receipts (i.e. recipes), but casual about the wedding arrangements.

Dear Polly,

As I was turning over some old mouldy volumes, that were laid upon a Shelf in a Closet of my Bed-chamber; one I found, after blowing the Dust from it with a Pair of Bellows, to be a Receipt Book, formerly, no doubt, belonging to some good old Lady of the Family. The Title Page (so much of it as the Rats had left) told us it was 'a Bouk off verry monny muckle vallyed Receipts bouth in Kookery and Physicks.' Upon one Page was 'To make Pye-Crust',—in another 'To make Wall-Crust',—'To make Tarts',—and at length 'To make Love'. 'This Receipt', says I, 'must be curious, I'll send it to Miss Howard next Post, let the way of making it be what it will.' [It was a long herbal recipe]. . . .

Then follow'd 'Another Receipt to make Love', which began 'Take two Sheep's Hearts, pierce them many times through with a Scewer to make them Tender, lay them upon a quick Fire, and then taking one Handful'—here Time with his long Teeth had gnattered away the remainder of this Leaf. At the Top of the next Page, begins 'To make an honest Man'. 'This is no new dish to me', says I, 'besides it is now quite old Fashioned; I won't read it'. Then follow'd 'To make a good Wife'. 'Pshaw', continued I, 'an acquaintance of mine, a young Lady of Lichfield, knows how to make this Dish better than any other Person in the World, and she has promised to treat me with it sometime', and thus in a Pett threw doun the Book, and would not read any more at that Time. If I should open it again tomorrow, whatever curious and useful receipts I shall meet

* These references at the ends of letters are to the books where a fuller text can most conveniently be found. The complete titles of the books, with dates, etc., are given in the book list, page 205.

with, my dear Polly may expect an account of them in another Letter.

I have the Pleasure of your last Letter, am glad to hear thy cold is gone, but do not see why it should keep you from the concert, because it was gone. . . .

I will certainly be with Thee on Wednesday evening, the Writings are at my House, and may be dispatched that night, and if a License takes up any Time (for I know nothing at all about these Things) I should be glad if Mr. Howard would order one, and by this means, dear Polly, we may have the Ceremony over next morning at eight o'clock, before any Body in Lichfield can know almost of my being come Home. If a License is to be had the Day before, I could wish it may be put off till late in the Evening, as the Voice of Fame makes such quick Dispatch with any News in so small a Place as Lichfield.—I think this is much the best scheme, for to stay a few Days after my Return could serve no Purpose, it would only make us more watch'd and teazed by the Eye and Tongue of Impertinence. . . .

(To Mary Howard, 24 December 1757. C. Darwin, *Life of Erasmus Darwin*, pp. 21–3)

Several letters to Boulton in the 1760s reveal the enthusiasm for technological improvement which led to the formation of the Lunar Society. The first letter is rather skittish:

As you are now become a sober plodding man of business I scarcely dare trouble you to do me a Favour in ye nicknachatery, alias philosophical way. . . .

Now if you like Florence Wine, I beg leave to make you a present of one Bottle, or two, if ye first don't answer, to drink success in Philosophy and Trade upon condition that you will procure me one of their Necks to be twisted into a little Hook according to ye copper plate on ye reverse of this Paper. It must be truly hermetically sealed, air tight, otherwise it will not answer my Experiment at all. . . .

I am extremely impatient for this new plaything! as I intend to fortell every Shower by it and make great medical Discoveries as far as relates to ye specific gravity of air: and from ye Quantity of Vapors.

(To M. Boulton, 1 July 1763. R. E. Schofield, *Lunar Society*, p. 29)

The second letter contains Darwin's enthusiastic proposal for a steam-carriage:

> As I was riding Home yesterday, I consid'd the Scheme of ye fiery Chariot—and ye longer I contemplated this favourite Idea, ye [more] practicable it appear'd to me. I shall lay my Thoughts before you, crude and indigested as they occur'd to me . . . as by those Hints you may be led into various Trains of thinking upon the Subject, and by that means (if any Hints can assist your Genius, which without Hints is above all others I am acquainted with) be more likely to improve or disapprove. And as I am quite mad of this Scheme, I beg you will not mention it, or show this Paper, to . . . any Body. . . .

After giving a detailed and quite plausible description of the mechanism for the steam-chariot, he continues:

> . . . if you think it feasible and will send me a Critique upon it, I will certainly . . . endeavour to build a Fiery Chariot; and if that answers get a patent. If you chuse to be Partner with me in ye Profit, and Expense and Trouble, let me know: as I am determined to execute it, if you approve of it.
> (To M. Boulton, 1763 or 1764. R. E. Schofield, *Lunar Society*, pp. 29–31)

Steam engines figure prominently in the next letter to Boulton.

> I am undone to know what observations Dr. Franklin supply'd you with about your Steam-Engine, besides giving you his approbation, and particularly to hear your final opinion, and Dr. Small's on ye important Question, whether Evaporation is at the Surface of boiling Water, or not?—or if it be at ye Surface of ye Vessel, exposed to ye Fire, which I rather suspect. . . .
> (To M. Boulton, 12 December 1765. R. E. Schofield, *Lunar Society*, p. 60)

Edgeworth's arrival at Lichfield in 1766 prompted this immediate call to Boulton:

> I have got with me a mechanical Friend, Mr. Edgeworth from Oxfordshire—The greatest Conjuror I ever saw—God send fair Weather, and pray come to my assistance, and prevail

on Dr. Small and Mrs. Boulton to attend you tomorrow Morning and we will re-convoy you to Birmingham on Monday, if ye D—l permit. He has ye principles of Nature in his palm, and moulds them as He pleases. Can take away polarity or give it to the Needle by rubbing it thrice on ye palm of his Hand.

And can see through two solid Oak Boards without Glasses! astonishing! diabolical!!! Pray tell Dr. Small He must come to see these Miracles.

(To M. Boulton, Summer 1766. R. E. Schofield, *Lunar Society*, p. 45)

A few months later Darwin was consulted about the murder of an illegitimate child by its mother. He replied:

Dear Sir,

I am sorry you should think it necessary to make any excuse for a Letter I this morning received from you. The Cause of Humanity needs no Apology to me. . . .

The Women that have committed this most unnatural crime, are real objects of our greatest Pity; their education has produced in them so much Modesty, or sense of Shame, that this artificial Passion overturns the very instincts of Nature!—what struggles must there be in their minds, what agonies!—at a Time when, after the Pains of Parturition, Nature has designed them the sweet Consolation of giving Suck to a little helpless Babe, that depends on them for its hourly existence!—Hence the cause of this most horrid crime is an excess of what is really a Virtue, of the Sense of Shame, or Modesty. Such is the Condition of human Nature! . . .

(To unknown man, 7 February 1767. C. Darwin, *Life of Erasmus Darwin*, pp. 28–9)

Darwin's first letter to Watt came shortly after his visit to Lichfield in the summer of 1767:

Now, my dear new friend, I first hope you are well and less hypochondriacal, and that Mrs. Watt and your child are well. The plan of your steam improvements I have religiously kept secret, but begin myself to see some difficulties in your execution which did not strike me when you were here. I have got another and another new hobby-horse since I saw you. I wish the Lord would send you to pass a week with me, and Mrs. Watt along with you;—a week, a month, a year! You promised to send me an instrument to draw landscapes with. If you ever move your place of residence for any long time from Glasgow, pray acquaint me. Adieu. Your friend,

E. Darwin

(To J. Watt, 18 August 1767. J. Muirhead, *James Watt*, I 15)

In 1768 Darwin was working on his horizontal windmill for Wedgwood's new factory at Etruria. But he advised Wedgwood (though without revealing Watt's secret) that steam might be better:

I should long ago have wrote to you, but waited to learn in what forwardness Mr. Watt's Fire-Engine was in. He has taken a Partner [Dr. Roebuck], and I can make no conjecture how soon you may be accommodated by Him with a Power so much more convenient than that of Wind. I will make packing Boxes and send you my model, that you may consult the Ingenious. I am of opinion it will be a powerful and a convenient Windmill, but would recommend steam to you if you can wait awhile. . . .

(To J. Wedgwood, 15 March 1769. E. Meteyard, *Life of Wedgwood*, II 31)

In fact Darwin's windmill was installed at Etruria nine years later, and worked well for many years (see Fig. 8).

In 1772 the great canal engineer, James Brindley, died of diabetes, which Darwin had diagnosed a few months before. Wedgwood's report of his death brought this reply.

Dear Wedgwood,

I did not return soon enough out of Derbyshire to answer your letter by yesterday's Post. Your second letter gave me great consolation about Mrs. Wedgewood, but gave me most sincere grief about Mr. Brindley, whom I have always esteemed to be a great genius, and whose loss is truly a public one. I don't believe he has left his equal. I think the various Navigations should erect him a monument in Westminster Abbey, and hope you will at the proper time give them this hint. . . . Any circumstances that you recollect of his life should be wrote down, and I will some time digest them into an Eulogium. These men should not die; this Nature denies, but their Memories are above her Malice.

(To J. Wedgwood, 30 September 1772. C. Darwin, *Life of Erasmus Darwin*, pp. 30–1)

Three of Darwin's letters to Benjamin Franklin have been preserved. Here is one of them:

Dear Sir,

I have inclosed a medico-philosophical paper which I should take it as a Favour if you will communicate to the royal Society, if you think it worthy a place in their Volume, otherwise must desire you to return it to the Writer.

I have another very curious paper containing Experiments on the Colours seen in the closed Eye after having gazed some time on luminous objects, which is not quite transcribed, but which I will also send you, if you think it is likely to be acceptable to the Society at this Time, but will otherwise let it lie by me another year. I hope you continue to enjoy . . . Health and that I shall sometime again have the pleasure of seeing you in Staffordshire, and am, dear Sr.

Your affectionate Friend,
Eras. Darwin

(To B. Franklin, 24 January 1774. L. Hussakov, *Science*, *43*, 774 (1916))

The 'very curious paper' mentioned is probably one on Ocular Spectra, later published under the name of Darwin's son Robert.

When Dr Small died in 1775, Darwin wrote inviting Dr Withering to take over the practice.

Dear Doctor,

I am at this moment returned from a melancholy scene, the death of a friend who was most dear to me, Dr. Small, of Birmingham, whose strength of reasoning, quickness of invention, learning in the discoveries of other men, and integrity of heart (which is worth them all), had no equal. Mr. Boulton suffers an inconceivable loss from the doctor's mechanical as well as medical abilities.

A person at Birmingham desired I would acquaint you with Dr. Small's death as soon as I could, but would not permit his name to be mentioned. . . . Now it occurred to me that if you should choose that situation [Dr. Small's practice] your philosophical taste would give you the friendship of Mr. Boulton,

which would operate all that for you which it did for Dr. Small. . . .
(To W. Withering, 25 February 1775. T. W. Peck and K. D. Wilkinson, *William Withering*, p. 1)

In the same year Watt was invited to work in Russia. Darwin was alarmed at the prospect.

Lord, how frightened I was when I heard a Russian bear had laid hold of you with his great paw, and was dragging you to Russia! Pray don't go if you can help it. Russia is like the den of Cacus: you see the footsteps of many beasts going thither, but of few returning. I hope your fire-engines will keep you here.
(To J. Watt, 29 March 1775. J. Muirhead, *James Watt*, II 84)

In quite a different vein we have Darwin's playful *billet doux* to Anna Seward, written as a letter from his Persian cat Mr Snow to her tabby Miss Po Felina.

Dear Miss Pussey,

As I sat, the other day, basking myself in the Dean's Walk, I saw you in your stately palace, washing your beautiful round face and elegantly brinded ears with your velvet paws, and whisking about with graceful sinuosity your meandering tail. That treacherous hedgehog, Cupid, concealed himself behind your tabby beauties, and darting one of his too well-aimed quills, pierced, O cruel imp! my fluttering heart.

Ever since that fatal hour have I watched, day and night, in my balcony, hoping that the stillness of the starlight evenings might induce you to take the air. . . .

Permit me this afternoon to lay at your divine feet the head of an enormous Norway Rat, which has even now stained my paws with its gore. If you will do me the honor to sing the following song . . . I will bring a band of catgut and catcall, to accompany you in chorus.

Cats I scorn who, sleek and fat,
Shiver at a Norway rat;
Rough and hardy, bold and free,
Be the cat that's made for me! . . .

Deign, most adorable charmer, to pur your assent to this my request, and believe me to be with the profoundest respect, your true admirer,

Snow.

(To Anna Seward, 7 September 1780. A. Seward, *Life of Dr Darwin*, pp. 134–7)

Anna's reply is longer but equally feline: her cat rejects Mr Snow because of his brutal habits.

O, should cat of Darwin prove
Foe to pity, foe to love!
Cat, that listens day by day
To mercy's mild and honied lay. . . .
(A. Seward, *Life of Dr Darwin*, p. 144)

The next three letters are all, in essence, apologies for not attending Lunar Society meetings. But rarely have apologies been so wittily offered.

Dear Boulton,

I am sorry the infernal Divinities who visit mankind with diseases, and are therefore at perpetual war with doctors, should have prevented my seeing all you great men at Soho today—Lord! What inventions, what wit, what rhetoric, metaphysical, mechanical, and pyrotechnical will be on the wing, bandy'd like a shuttlecock from one to another of your troop of philosophers! While poor I, I by myself I, imprizon'd in a post chaise, am joggled, and jostled, and bump'd, and bruised along the King's high road, to make war upon a pox or a fever!

Now had I pack'd up a print of Prince pretty-man, for you, and a book on opake colours for the society, in vain!

(To M. Boulton, 5 April 1778. R. E. Schofield, *Lunar Society*, pp. 143–4)

When Watt invited him to a meeting in January 1781 he had to make a similar excuse:

You know there is a perpetual war carried on between the devil and all holy men. Sometimes one prevails in an odd skirmish or so, and sometimes the other. Now, you must know this said devil has played me a slippery trick, and, I fear, prevented me from coming to join the holy men at your house, by sending the measles with peripneumony amongst nine beautiful children of Lord Paget's. For I must suppose it is a work of the devil! Surely the Lord could never think of amusing himself by setting nine innocent little animals to cough their hearts up? Pray ask your learned society if this partial evil contributes to

> any public good? . . . Pray inquire of your philosophers, and rescue me from Manichaeism.
>
> As to material philosophy, I can tell you some secrets in return for yours; viz., that . . . water is composed of aqueous gas, which is displaced from its earth by oil of vitriol. . . .
> (To J. Watt, 6 January 1781. J. Muirhead, *James Watt*, II 124)

At the end of this letter Darwin seems to be suggesting that water is not an element and that hydrogen is one of its constituents, a speculation apparently taken up by Priestley a month or two later, in experiments which were to lead Watt in 1783 to his correct suggestion about the composition of water.[18]

By 1782 Darwin was living near Derby and too far away to come to the Lunar meetings very often.

> I am here cut off from the milk of science, which flows in such redundant streams from your learned lunatics; which, I can assure you is a very great regret to me. . . . Pray if you think of it, make my devoirs to the learned Insane of Your society. . . .
> (To M. Boulton, 28 December 1782. R. E. Schofield, *Lunar Society*, p. 204)

To make up for his absence from Lunar meetings, Darwin founded the Derby Philosophical Society, as he explained to Boulton in 1783:

> We have established an infant philosophical Society at Derby, but do not presume to compare it to your well-grown gigantic philosophers at Birmingham. . . . I wish you would bring a party of your Society and hold one Moon at our house. N.B. our Society intend to eclipse the moon on the 18th of this month, pray don't you counteract our conjurations.
> (To M. Boulton, 4 March 1783. R. E. Schofield, *Lunar Society*, p. 236)

In 1784 Darwin began negotiations with the publisher Joseph Johnson about his poem *The Loves of the Plants*. Darwin's requirements were quite exacting:

> (1) I would not have my name affixed to this work on any account, as I think it would be injurious to me in my medical practice, as it has been to all other physicians who have pub-

lished poetry. (2) I would not wish to part with the entire copy-right. . . .

What I would propose to you is that we publish one edition consisting of 500 copies . . . and afterwards, if you think proper, to publish another 500 copies . . . but that after the publication of this 1000 copies, I shall again be at liberty to do what I please with the work. . . .

(To J. Johnson, 23 May 1784. R. E. Schofield, *Lunar Society*, pp. 204–5)

In 1787 Darwin helped his son Robert with the paper on Ocular Spectra and canvassed support for his election as a Fellow of the Royal Society, as shown by this letter to Wedgwood:

When I want anything to be done . . . I look out for a man who has the most business of his own; for if I can prevail on him to undertake it, it is sure to be done soon and well! Hence I apply to you, who have more to do of your own than any other man I have the pleasure to be acquainted with. My son, Dr. Robert Waring Darwin, of Shrewsbury, I wish to be a member of the Royal Society, as it would be a feather in his cap, and might encourage him in philosophical pursuits; and I flatter myself he will make an useful member of that ingenious society. Now, I am unacquainted with the necessary form, and *if it be not either too troublesome or Disagreeable to you*, I should be much obliged to you to hang up his name, and take what steps are necessary.

(To J. Wedgwood, 8 May 1787. E. Meteyard, *A Group of Englishmen*, pp. 253–4)

The chemists of the Lunar Society—Priestley, Watt and Keir—stubbornly clung to the phlogiston theory of air into the 1790s.* But Darwin saw its fallacies and accepted the 'French chemistry', as expounded by Fourcroy, as early as 1788, when he wrote bluntly to Watt:

*According to the phlogiston theory, burning substances gave out 'phlogiston': since in fact they combine with oxygen and are heavier after combustion than before, the 'phlogiston' must have had negative weight. However, anyone who holds an eccentric belief can explain away inconvenient facts, and the phlogistians were no exception.

Pray read the 40 first pages of the Introduction to Fourcroy's Elements, and tell me if the facts are in general true.—If they be, the theory holds them nicely together. When steam is passed through red hot iron scrapings, if the water be not decomposed, whence comes the vital air, which unites to the iron? does the water vanish, or is it annihilated? pray explain this experiment.
(To J. Watt, 18 November 1788. R. E. Schofield, *Lunar Society*, p. 293)

We do not have Watt's reply, but Darwin wrote similar letters to Keir, and one of Keir's replies, written in March 1790, serves to show the 'state of play' in this dispute.

The proof of the composition of water, though necessary for Lavoisier's system, does not at all disprove the theory of phlogiston. . . . You are such an infidel in religion that you cannot believe in transubstantiation, yet you can believe that apples and pears, hay and oats, bread and wine, sugar, oil, and vinegar, are nothing but water and charcoal, and that it is a great improvement in language to call all these things by one word, oxyde hydro-carbonneux. . . .
(J. Keir to Darwin, 15 March 1790. J. K. Moilliet, *James Keir*, p. 111)

In 1789 the abolition of the slave-trade was still eighteen years ahead, but Darwin and Wedgwood were already working hard for abolition, with propaganda of all kinds. Wedgwood's cameo of a slave in chains, with the motto 'Am I not a man and a brother?', was widely circulated, and Darwin was always on the alert for useful propaganda:

I have just heard that there are muzzles or gags made at Birmingham for the slaves in our islands. If this be true, and such an instrument could be exhibited by a speaker in the House of Commons, it might have a great effect. Could not one of their long whips or wire tails be also procured and exhibited? But an instrument of torture of our own manufacture would have a greater effect, I dare say.
(To J. Wedgwood, 13 April 1789. C. Darwin, *Life of Erasmus Darwin*, p. 46)

The outbreak of the French Revolution in July 1789 was welcomed by all the Lunar group. Wedgwood wrote to Darwin,

I know you will rejoice with me in the glorious revolution which has taken place in France. The Politicians tell me that as a manufacturer I shall be ruined if France has her liberty. . . . I [do not] see that the happiness of one nation includes in it the misery of its next neighbour.

(J. Wedgwood to Darwin, July 1789. R. E. Schofield, *Lunar Society*, p. 358)

And Darwin asked Watt:

Do you not congratulate your grandchildren on the dawn of universal liberty! I feel myself becoming all french in both chemistry and politics.

(To J. Watt, 19 January 1790. R. E. Schofield, *Lunar Society*, p. 358)

The next letter brings us from public affairs to private grief. Edgeworth's fifteen-year-old daughter Honora died in 1790. Darwin wrote:

Dear Edgeworth,

I much condole with you on your late loss. I know how to feel for your misfortune. The little Tale you sent me is a prodigy, written by so young a person, with such elegance of imagination. [Honora had written a tale called 'Rivuletta'.]

Nil admirari may be a means to escape misery, but not to procure happiness. There is not much to be had in this world—we *expect* too much.

I have had my loss also! The letter of Sulpitius to Cicero is fine eloquence, but comes not to the heart; it tugs, but does not draw the arrow. Pains and diseases of the mind are only cured by Time. Reason but skins the wound, which is perpetually liable to fester again.

(To R. L. Edgeworth, early 1790. Edgeworth, *Memoirs*, p. 323)

Darwin was always on good terms with his son Robert, and since Robert lived about fifty miles away, in Shrewsbury, many long letters passed between them. Some of these are purely medical, but others are of more general interest.

Dear Robert,

I am sorry to hear you say you have many enemies, and one enemy often does much harm. The best way, when any little slander is told one, is never to make any piquant or angry

> answer; as the person who tells you what another says against you, always tells them in return what you say of them. I used to make it a rule always to receive all such information very coolly, and never to say anything biting against them which could go back again; and by these means many who were once adverse to me, in time became friendly. Dr. Small always went and drank tea with those who he heard had spoken against him; and it is best to show a little attention at public assemblies to those who dislike one; and it generally conciliates them.
> (To R. W. Darwin, 19 April 1789. C. Darwin, *Life of Erasmus Darwin*, p. 37)

Good advice was also forthcoming on the question of making money.

> As to fees, if your business pays you well on the whole, I would not be uneasy about making absolutely the most of it. To live comfortably all one's life, is better than to make a very large fortune towards the end of it.
> (To R. W. Darwin, 7 February 1792. C. Darwin, *Life of Erasmus Darwin*, p. 65)

In about 1793, when Erasmus was sixty-two, Robert urged him to retire. He replied with characteristic bluntness,

> it is a dangerous experiment, and generally ends either in drunkenness or hypochondriacism. Thus I reason, one must do something (so country squires fox-hunt), otherwise one grows weary of life, and becomes a prey to ennui. Therefore one may as well do something advantageous to oneself and friends or to mankind, as employ oneself in cards or other things equally insignificant.
> (To R. W. Darwin, 1793? C. Darwin, *Life of Erasmus Darwin*, p. 52)

Darwin's letters to his son are much less skittish than those to his friends in the Lunar Society: it was the Lunar tradition to be light-hearted at every opportunity.

Jenner's discovery of vaccination came right at the end of Darwin's medical career, but as usual he was ready to welcome the new idea and wanted to legitimize it by links with older customs:

> Your discovery of preventing the dreadful havoc made

among mankind by the small-pox, by introducing into the system so mild a disease as the vaccine inoculation produces, may in time eradicate the small-pox from all civilized countries, and this especially: as by the testimony of innumerable instances the vaccine disease is so favourable to young children, that in a little time it may occur that the christening and vaccination of children may always be performed on the same day. (To E. Jenner, 24 February 1802. J. Baron, *Life of Jenner*, II 541)

Darwin's last letter was begun the day before he died and was never finished.

Dear Edgeworth. . . .

We have all been now removed from Derby about a fortnight, to the Priory, and all of us like our change of situation. We have a pleasant house, a good garden, ponds full of fish, and a pleasing valley somewhat like Shenstone's—deep, umbrageous, and with a talkative stream running down it. Our house is near the top of the valley, well screened by hills from the east, and north, and open to the south, where, at four miles distance, we see Derby tower.

Four or more strong springs rise near the house, and have formed the valley, which, like that of Petrarch, may be called *Val chiusa*, as it begins, or is shut, at the situation of the house. I hope you like the description, and hope farther, that yourself and any part of your family will sometime do us the pleasure of a visit. . . .

(To R. L. Edgeworth, 17 April 1802. Edgeworth, *Memoirs*, p. 396)

3 Eye-witness Reports

Report me and my cause aright.

SHAKESPEARE; *Hamlet*

DARWIN'S letters leave us with a strong impression of his personality: but it is also worth hearing what his contemporaries said about him.

Anna Seward must be allowed to fire the first shots, for she tells us about Darwin at excessive length in her 430-page biography. Here is her account of what he looked like in his early years at Lichfield:

> He was somewhat above the middle size, his form athletic, and inclined to corpulence; his limbs too heavy for exact proportion. The traces of a severe small-pox; features, and countenance, which, when they were not animated by social pleasure, were rather saturnine than sprightly; a stoop in the shoulders, and the then professional appendage, a large full-bottomed wig, gave, at that early period of life, an appearance of nearly twice the years he bore. Florid health, and the earnest of good humour, a sunny smile, on entering a room, and on first accosting his friends, rendered, in his youth, that exterior agreeable, to which beauty and symmetry had not been propitious.
>
> He stammered extremely; but whatever he said, whether gravely or in jest, was always well worth waiting for. . . .
>
> (A. Seward, *Life of Dr Darwin*, pp. 1–2)

Anna's style is so extraordinarily long-drawn out that we can only savour one of the many episodes she relates, the story of Darwin's second marriage. After quoting his verses to Mrs Pole, Anna continues as follows.

> During the course of the year 1780, died Colonel Pole. Dr. Darwin, more fortunate than Petrarch, whose destiny his own had resembled in poetic endowment and hopeless love, then saw his adored Laura free, and himself at liberty to court her favor. . . . He soon, however, saw her surrounded by

rivals, whose time of life had nearer parity with her own, yet in its summer bloom, while his age nearly approached its half century; whose fortunes were affluent and patrimonial; while his were professional; who were jocund bachelors, while he had children for whom he must provide.

Colonel Pole had numbered twice the years of his fair wife. . . .

Mrs. Pole, it has already been remarked, had much vivacity and sportive humor, with very engaging frankness of temper and manners. Early in her widowhood she was rallied in a large company upon Dr. Darwin's passion for her, and was asked what she would do with her captive philosopher. 'He is not very fond of churches, I believe, and if he would go there for my sake, I shall scarcely follow him. He is too old for me.' —'Nay, madam, what are fifteen years on the right side?' She replied, with an arch smile, 'I have had so *much* of that right side!'

The confession was thought inauspicious to the Doctor's hopes; but it did not prove so; the triumph of intellect was complete . . . she took Dr. Darwin for her husband. Darwin, never handsome, or personally graceful, with extremely impeded utterance; with hard features on a rough surface; older much in appearance than in reality; lame and clumsy!—and this, when half the wealthy youth of Derbyshire were said to have disputed the prize with him.

But it was not without some stipulations. . . . She had taken a dislike to Lichfield, and decidedly said, nothing could induce her to live there.

(A. Seward, *Life of Dr Darwin*, pp. 147–151)

Many of the family stories about Darwin appear in the excellent biographical sketch written by his grandson Charles, who remembered the anecdotes he had heard from his own father, Erasmus's son Robert.

A young man once asked him in, as he thought, an offensive manner, whether he did not find stammering very inconvenient. He answered, 'No, Sir, it gives me time for reflection, and saves me from asking impertinent questions'. . . .

When he wished to make himself disagreeable for any good cause, he was well able to do so. Lady —— married a widower, and became so jealous of his former wife that she cut and spoiled her picture, which hung up in one of the rooms. The husband, fearing that his young wife was becoming insane, was greatly alarmed, and sent for Dr. Darwin. When he arrived he told her in the plainest manner many unpleasant truths, amongst others that the former wife was infinitely her superior in every respect, including beauty. The poor lady was astonished at being thus treated, and could never afterwards endure his name. He told the husband if she again behaved oddly, to hint that he would be sent for. The plan succeeded perfectly, and she ever afterwards restrained herself. . . .

My father was much separated from Dr. Darwin after early life, so that he remembered few of his remarks, but he used to quote one saying as very true: 'that the world was not governed by the clever men, but by the active and energetic'. He used also to quote another saying, that 'common sense would be improving, when men left off wearing as much flour on their heads as would make a pudding; when women left off wearing rings in their ears, like savages wear nose rings; and when fire-grates were no longer made of polished steel'. . . .

The gratitude of the poor to him was shown on two occasions in a strange manner. . . . The second story is, that as the doctor was riding at night on the road to Nottingham a man on horseback passed him, to whom he said good night. As the man soon slackened his pace, Dr. Darwin was forced to pass him, and again spoke, but neither time did the man give any answer. A few nights afterwards a traveller was robbed at nearly the same spot by a man who, from the description, appeared to be the same. It is added that my grandfather out of curiosity visited the robber in prison, who owned that he had intended to rob him, but added: 'I thought it was you, and when you spoke I was sure of it. You saved my life many years ago, and nothing could make me rob you.'

(C. Darwin, *Life of Erasmus Darwin*, pp. 40, 42–3, 63–5)

His reputation for treating poor patients free earned a mention in one of Sir Brooke Boothby's sonnets:

Darwin, thine whose ever-open door
Draws, like Bethesda's pool, the suffering poor;
Where some fit cure the wretched all obtain
Relieved at once from poverty and pain.

(C. Darwin, *Life of Erasmus Darwin*, p. 63)

After Darwin's death, James Keir, a friend for nearly fifty years, wrote as follows to Robert Darwin:

> I think all those who knew him will allow that sympathy and benevolence were the most striking features. He felt very sensibly for others, and, from his knowledge of human nature, he entered into their feelings and sufferings in the different circumstances of their constitution, character, health, sickness, and prejudice. In benevolence, he thought that almost all virtue consisted. He despised the monkish abstinences and the hypocritical pretensions which so often impose on the world. The communication of happiness and the relief of misery were by him held as the only standard of moral merit. Though he extended his humanity to every sentient being, it was not like that of some philosophers, so diffused as to be of no effect; but his affection was there warmest where it could be of most service to his family and his friends, who will long remember the constancy of his attachment and his zeal for their welfare. . . . Your father did indeed retain more of his original character than almost any man I have known, excepting, perhaps, Mr. Day. Indeed the originality of character in both these men was too strong to give way to the example of others. . . .
>
> The works of your father are a more faithful monument and more true mirror of his mind than can be said of those of most authors. . . .
>
> (J. Keir, letter to R. W. Darwin, 12 May 1802. C. Darwin, *Life of Erasmus Darwin*, pp. 35–6, 49, 68)

James Watt's tributes to Darwin were short and to the point:

> For my part, it will be my pride, while I live, that I have enjoyed the friendship of such a man. . . . He was almost my most ancient acquaintance and friend in England, I having been intimate with him for thirty-four years, and on many occasions much indebted to his good offices.
>
> (J. Watt, letters to R. W. Darwin, 24 April 1802 and to R. Muirhead, 3 May 1802. J. Muirhead, *James Watt*, II 279)

When Edgeworth first visited Darwin, the Doctor was out on his rounds. But his return was dramatic:

> When supper was nearly finished, a loud rapping at the door

announced the Doctor. There was a bustle in the hall, which made Mrs. Darwin get up and go to the door. Upon her exclaiming that they were bringing in a dead man, I went to the hall: I saw some persons, directed by one whom I guessed to be Doctor Darwin, carrying a man who appeared motionless.

'He is not dead', said Doctor Darwin. 'He is only dead drunk. I found him', continued the Doctor, 'nearly suffocated in a ditch; I had him lifted into my carriage, and brought hither, that we might take care of him to-night.'

Candles came, and what was the surprise of the Doctor, and of Mrs. Darwin, to find that the person whom he had saved was Mrs. Darwin's brother! who, for the first time in his life, as I was assured, had been intoxicated in this manner, and who would undoubtedly have perished, had it not been for Doctor Darwin's humanity.

During this scene I had time to survey my new friend, Doctor Darwin. He was a large man, fat, and rather clumsy; but intelligence and benevolence were painted in his countenance: he had a considerable impediment in his speech, a defect which is in general painful to others; but the Doctor repaid his auditors so well for making them wait for his wit or his knowledge, that he seldom found them impatient.

(R. L. and Maria Edgeworth, *Memoirs*, pp. 102–3)

Edgeworth's daughter Maria was with him on several of his later visits to Darwin, and she describes his attacks on alcohol:

Dr. Darwin . . . had a strong belief that almost all the distempers of the higher classes of people arise from drinking, in some form or other, too much vinous spirit. . . . This opinion he supported in his writings with the force of his eloquence and reason; and still more in conversation, by all those powers of wit, satire, and peculiar humour, which never appeared fully to the public in his works, but which gained him strong ascendancy in private society. During his life-time, he almost banished wine from the tables of the rich of his acquaintance; and persuaded most of the gentry in his own and the neighbouring counties, to become water-drinkers. . . . When he heard that my father was bilious, he suspected that this must be the consequence of his having . . . indulged too freely in drinking. His letter, I remember, contained a humorous philippic against 'vinous potation', and concluded with—'Farewell, my dear friend. God keep you from whiskey—if he can.'

To anyone who knew my father, this must seem a laughable suspicion, for he was famous, or in those days I may say, infamous, in Ireland, for his temperance.

(R. L. and Maria Edgeworth, *Memoirs*, pp. 299–300)

Samuel Galton's daughter Mary Anne, later Mrs SchimmelPenninck, was deeply religious and a notorious mischief-maker: even the kindest of her relatives, her nephew Sir Francis Galton, noted that 'she broke off *eleven* marriages'. Mary Anne was ten years old when Darwin came to treat her mother, and she took an instant dislike to him. Her description of Darwin was written down sixty years later, and is so vivid that it deserves to be quoted, even though quite unreliable in its details.

It was in the autumn of 1788 that the celebrated Dr. Darwin first came to see my mother. It was in the latter part of the morning that a carriage drove up to our door of the description then called a sulky because calculated to hold one person only. The carriage was worn, and bespattered with mud. Lashed on the place appropriated to the boot in ordinary carriages was a large pail for the purpose of watering the horses, together with some hay and oats beside it. In the top of the carriage was a skylight, with an awning which could at pleasure be drawn over; this was for the purpose of giving light to the doctor, who wrote most of his works on scraps of paper with a pencil as he travelled. The front of the carriage within was occupied by a receptacle for writing-paper and pencils, likewise for a knife, fork and spoon; on one side was a pile of books reaching from the floor to nearly the front window of the carriage; on the other, a hamper containing fruit and sweetmeats, cream and sugar, the greater part of which, however, was demolished during the time the carriage traversed the forty miles which separated Derby from Barr. We all hastened to the parlour window to see Dr. Darwin, of whom we had heard so much, and whom I was prepared to honour and venerate, in no common degree, as the restorer of my mother's health. What then was my astonishment at beholding him as he slowly got out of the carriage! His figure was vast and massive, his head was almost buried on his shoulders, and he wore a scratch wig, as it was then called, tied up in a little bob-tail behind. A habit of stammering made the closest attention necessary, in order to understand what he said. Meanwhile, amidst all this, the doctor's eye was deeply sagacious, the most so I think of any eye I remember ever to have seen; and I can conceive that no patient consulted Dr. Darwin who, so far as intelligence was concerned, was not inspired with confidence in beholding him; his observation was most keen; he constantly detected disease, from his sagacious observation of symptoms apparently so slight as to be unobserved by other doctors. His horror of fermented liquors, and his belief in the advantages both of eating largely, and eating an almost immeasurable abundance of sweet things, was well known to all his friends; and we had on this occasion, as indeed was the custom whenever he came, a luncheon-table set out with hothouse fruits, and West India sweetmeats, clotted cream, Stilton cheese, etc. When the whole party were settled at table and I had lost the fear that the doctor would speak to me, and when, by dint of attention, I could manage to understand what he said, I was astonished at his wit, his anecdotes, and most entertaining conversation. . . .

In the beginning of 1789 my mother was again far from well, and my father sent for Dr. Darwin. . . . His whole conversation, I remember, on that occasion was characterized by the merriment and so-called wit which aimed its perpetual shafts against those holy truths which, imperfectly though I yet knew them, afforded me the only comfort in distress which I had ever experienced. . . .

Dr. Darwin often used to say, 'Man is an eating animal, a drinking animal, and a sleeping animal, and one placed in a material world, which alone furnishes all the human animal can desire. He is gifted besides with knowing faculties, practically to explore and to apply the resources of this world to his use. These are realities. All else is nothing; conscience and sentiment are mere figments of the imagination. Man has but five gates of knowledge, the five senses; he can know nothing but through them; all else is a vain fancy, and as for the being of a God, the existence of a soul, or a world to come, who can know anything about them? Depend upon it, my dear madam, these are only the bugbears by which men of sense govern fools.'
(*Life of Mary Anne SchimmelPenninck*, I 151–3, 242)

If she really could remember all this verbatim after sixty years, Darwin must have made a lasting impression on her. However, she is obviously exaggerating at times.

Darwin's religious views also offended Coleridge, who visited him in 1796, and wrote:

> Derby is full of curiosities, the cotton, the silk mills, Wright, the painter, and Dr. Darwin, the everything, except the Christian! Dr. Darwin possesses, perhaps, a greater range of knowledge than any other man in Europe, and is the most inventive of philosophical men. He thinks in a *new* train on all subjects except religion. . . . Dr. Darwin is an extraordinary man, and received me very courteously—He had heard that I was a Unitarian and bantered incessantly on the subject of Religion. . . . When he talks on any other subject he is a wonderfully entertaining and instructive old man. . . .
>
> I absolutely nauseate Darwin's poem.
>
> (Coleridge, *Letters*, I 177, 178, 216)

Darwin's verses may have nauseated Coleridge, but they pleased everyone else, as Coleridge himself remarked in a letter to Godwin in 1801:

> I have myself met with persons who professed themselves idolatrous admirers of Milton, and yet declared it to be their opinion that Dr. Darwin was as great a poet. (Coleridge, *Letters*, II 738)

Horace Walpole, usually a difficult man to satisfy, was among the enthusiasts:

> Dr. Darwin has destroyed my admiration for any poetry but his own. . . . The twelve verses that by miracle describe and comprehend the creation of the universe out of chaos, are in my opinion the most sublime passage in any author, or in any of the few languages with which I am acquainted. There are a thousand other verses most charming, or indeed all are so, crowded with most poetic imagery, gorgeous epithets and style. (H. Walpole, *Letters*, XIV 41, 110)

This response is confirmed by dozens of others. Even Mrs SchimmelPenninck remembered the great pleasure the *Botanic Garden* gave her family, an impression which, 'though at the distance of sixty years, I shall never forget'.

The great engineer Thomas Telford called it 'a very wonderful and masterly performance'. James Keir, after reading it in manuscript in 1787, wrote, 'I am confirmed in the opinion I always had, that you would have been the first Poet of the Kingdom if you had not suppressed your talent'. (J. K. Moilliet, *James Keir*, p. 78). Edgeworth was even more enthusiastic:

> I have felt such continued, such increasing admiration in reading the 'Loves of the Plants', that I dare not express my enthusiasm, lest you should suspect me of that tendency to exaggeration, with which you used to charge me. I may, however, without wounding your delicacy, say that it has silenced for ever the complaints of poets who lament that Homer, Milton, Shakespeare, and a few classics, had left nothing new to describe, and that elegant imitation of imitations was all that could be expected in modern poetry.
>
> I have seen nobody since it has been published, except my own family; and amongst my domestic critics, who are not readily pleased, I hear nothing but praise and congratulation. (R. L. and Maria Edgeworth, *Memoirs*, p. 324)

There were many tributes in verse too, a typical one being William Cowper's 'Lines Addressed to Dr Darwin':

Two poets* (poets by report
Not oft so well agree)
Sweet Harmonist of Flora's court!
Conspire to honour thee.

They best can judge a poet's worth,
Who oft themselves have known
The pangs of a poetic birth
By labours of their own.

We, therefore, pleas'd, extol thy song,
Though various, yet complete,
Rich in embellishment, as strong,
And learn'd, as it is sweet. . . .

* Hayley was the second one.

And deem the bard, whoe'er he be,
  And howsoever known,
Who would not twine a wreath for thee,
  Unworthy of his own.

(W. Cowper, *Poetical Works* (O.U.P., 1911) p. 418)

In June 1802 Darwin received a candid seven-page obituary notice in the *Monthly Magazine*. Two items in this notice would have received the full approval of Mrs Schimmel-Penninck:

> He was in the habit of eating a large quantity of food, and his stomach possessed a strong power of digestion. . . .
> There are reasons for suspecting that Dr. Darwin was not a believer in Divine Revelation. . . .

The obituarist was also discerning in his judgment of Darwin's poems:

> No man, perhaps, was ever happier in the selection and composition of his epithets, had a more imperial command of words, or could elucidate with such accuracy and elegance the most complex and intricate machinery. . . .
> (*Monthly Magazine*, Vol. XIII, pp. 459, 463, 462)

Darwin's great reputation for about thirty years after his death is apparent in many casual references. For example, when in 1831 the Hydrographer of the Navy found a naturalist to go on the voyage of the *Beagle*—it was of course Charles Darwin—he wrote to FitzRoy, Captain of the *Beagle*, as follows:

> My dear Sir, I believe my friend Mr. Peacock of Trinity College Cambridge has succeeded in getting a 'Savant' for you —a Mr. Darwin grandson of the well known philosopher and poet. . . .
> (G. de Beer, *Charles Darwin*, p. 23)

Elizabeth Anne Galton, Darwin's grand-daughter, stated that 'in London and at school we were always introduced as Grand-daughters of Dr. Erasmus Darwin, and made much of in consequence'. She also remarked:

When they heard that my mother was daughter to Dr. Erasmus Darwin it was with difficulty my father could make them take a fee. Sir Astley Cooper almost embraced my Mother, he was so pleased to see a daughter of Dr. Darwin.
(Elizabeth Anne Wheler, *Memorials of my Life* (unpublished), pp. 121, 62)

4 A Model for Young Ladies

Three little maids from school are we.
W. S. GILBERT; *The Mikado*

AS A pleasant introduction to Darwin's weightier works, we have his little book *A Plan for the Conduct of Female Education in Boarding Schools* (1797). In about 1793 he arranged for his daughters Susan and Mary Parker to set up a boarding school at Ashbourne, and they asked him to help with the curriculum. They found his advice very useful and persuaded him to publish his ideas. In this book Darwin is on his best behaviour and writes nothing that would scandalize the parents of prospective pupils. He still manages to be quite radical, however, and his book has earned a distinctive place in the history of girls' education in England.[19] Darwin was totally opposed to the conventional idea that girls should be encouraged to become empty-headed simpering misses, and he went a long way towards the modern idea of equal treatment for boys and girls, as well as anticipating many subsequent reforms in the curriculum. His ideas had some practical effect too, because many of the leading families in the Midlands sent their girls to the Ashbourne school.

Even in the generalities at the beginning of the book, Darwin departs from convention by emphasizing the importance of physical education.

> The advantages of a good education consist in uniting health and agility of body with cheerfulness and activity of mind; in superadding graceful movements to the former, and agreeable tastes to the latter; and in the acquirement of the rudiments of such arts and sciences as may amuse ourselves, or gain us the esteem of others; with a strict attention to the culture of morality and religion. (p. 10)*

* In this chapter, and in Chapters 5–7, I have modernized the texts of Darwin's books, by eliminating most of the archaisms of spelling and punctuation: e.g. the possessive *it's* has been printed as *its*, and the comma which precedes nearly all eighteenth-century

Although Darwin believes that 'the female character should possess the mild and retiring virtues rather than the bold and dazzling ones', he half-contradicts this view in the next paragraph.

> There are however situations in single life, in which, after the completion of their school-education, ladies may cultivate to any extent the fine arts or the sciences for their amusement or instruction. And there are situations in a married state which may call forth all the energies of the mind in the care, education, or provision for a family; which the inactivity, folly, or death of a husband may render necessary. Hence if to softness of manners, complacency of countenance, gentle unhurried motion, with a voice clear and yet tender, the charms which enchant all hearts! can be superadded internal strength and activity of mind, capable to transact the business or combat the evils of life; with a due sense of moral and religious obligation; all is obtained which education can supply; the female character becomes complete, excites our love, and commands our admiration. (pp. 10–11)

After these generalities Darwin gets down to detail, in thirty-nine sections on specific topics, beginning with music and dancing, which he thinks are usually given too much attention. On the subject of writing he is eminently practical:

> A proper manner of holding the pen, or pencil, or needle, with an easy but graceful attitude of the person, and an agreeable moderate attention of the countenance, should first be taught; for which purposes an inclined desk has many advantages over an horizontal table for the books, or working frames, as the body is thence less bent forwards, and the light in general situations more vividly reflected to the eye.
>
> If the desk be sixteen inches broad, the furthermost edge of it should rise about 3½ inches from the horizontal line; which produces the most convenient inclination. . . . (pp. 15–16)

The idea of sloping desks has been widely adopted in schools,

---

relative clauses has been dropped when it conflicts with modern usage. I have also corrected some lapses in punctuation and grammar.

though the makers of office desks still retain a supinely horizontal attitude.

Darwin is rather scathing about grammar, and in his section on Languages he anticipates later reforms in the curriculum:

> The necessity of learning some ancient or foreign languages imposes a laborious task on the youth of both sexes; which consumes years of their precious time, which might otherwise be employed in the acquisition of sciences. The difficulty of obtaining a competent knowledge of the Greek or Latin language is many times greater than that of obtaining any modern one; as may be deduced from the innumerable changes of the termination of their nouns, adjectives and verbs, which to a beginner are all so many new words. And as the works of the best writers in these languages have been translated into our own, it is less necessary in the education of ladies to expend so much time and labour in acquiring them. But as the French and Italian are less difficult to learn, and contain new books of taste and knowledge, which are yearly published in this age of literature; and as they are convenient for conversing with foreigners who come hither, or in our travelling into other countries; and lastly, as they are agreeable as well as fashionable studies; the pupils of boarding schools should be encouraged to attain one or both of them. (pp. 17–18)

It was not only for girls that Darwin recommended modern languages in preference to ancient, as appears from one of Wedgwood's letters:

> I had some talk with Dr. Darwin upon my plan of curtailing the education of my boys in order to establish their health and give the more strength to their constitutions. He approved of the idea, and said he thought it a very idle waste of time for any boys intended for trade to learn Latin, as they seldom learnt it to any tolerable degree of perfection, or retained what they learnt. Besides they did not want it, and the time would be much better bestowed in making themselves perfect in French and accounts.[20]

Darwin's recommendations on other subjects are much as would be expected, but he strongly objects to the idea that women should remain forever feather-brained:

> I cannot conclude this section on polite learning without mentioning that some illiterate men have condemned the cultivation of the minds of the female sex, and have called such in ridicule 'learned ladies'; as if it was a reproach to render themselves agreeable and useful. Where affectation is joined with learning, it becomes pedantry, but this belongs oftener to the ignorant than to the cultivated. (pp. 39–40)

Darwin is again well ahead of his time in suggesting that young ladies should learn scientific subjects, including practical work ('experimental philosophy'), and visit factories ('manufactories').

> The various arts and manufactories which adorn and enrich this country should occasionally be shown and explained to young persons, as so many ingenious parts of experimental philosophy; as well as from their immediately contributing to the convenience of life, and to the wealth of the nations which have invented or established them. Of these are the cotton works on the river Derwent in Derbyshire; the potteries in Staffordshire; the iron-foundries of Coalbrooke Dale in Shropshire; the manufactories of Birmingham, Manchester, Nottingham; but these are not in the province of a boarding school, and might be advantageously exhibited to young ladies by their parents in the summer vacations. (p. 43)

Darwin also suggests that girls should learn shorthand, an idea which has flowered more abundantly than he could ever have foreseen.

After disposing of the academic subjects, Darwin turns to morals. Today, with the decline of religious belief, we need a secular morality for schools; and this is just what Darwin offers, for he makes only a few guarded references to religion. He is at his best on Compassion:

> A sympathy with the pains and pleasures of others is the foundation of all our social virtues. 'Do as you would be done by' is a precept which descended from heaven. Whoever feels pain himself when he sees others affected with it, will not only never be liable to give pain, but will always be inclined to relieve it. The lady who possesses this christian virtue of compassion cannot but be a good daughter, a good wife, and a good

> mother, that is, an amiable character in every department of life. . . .
>
> This compassion, or sympathy with the pains of others, ought also to extend to the brute creation, as far as our necessities will admit; for we cannot exist long without the destruction of other animal or vegetable beings either in their mature or embryon state. Such is the condition of mortality, that the first law of nature is 'eat, or be eaten'. Hence for the preservation of our existence we may be supposed to have a natural right to kill those brute creatures which we want to eat, or which want to eat us; but to destroy even insects wantonly shows an unreflecting mind, or a depraved heart.
>
> A young gentleman once assured me that he had lately fallen in love with a young lady; but, on their walking out one evening in summer, she took two or three steps out of her way on the gravel walk to tread upon an insect; and that afterwards whenever the idea of her came into his mind, it was attended with this picture of active cruelty; till that of the lady ceased to be agreeable, and he relinquished his design of courtship. (pp. 46–7; 48)

This cautionary tale should be enough to make the giddiest girl think twice. But, if not, Darwin has other plans in reserve.

> There should also be a plan in schools to promote the habit as well as the principle of benevolence; each young lady might occasionally contribute a small sum on seeing a needy naked child, to purchase flannel or coarse linen for clothes, which they might learn to cut out, and to make up themselves; and thus the practice of industry might be united with that of liberality. (p. 49)

Darwin also commends fortitude, and is scornful of girls who dissolve into tears at the slightest provocation. 'When a young lady screams through the whole gamut at the sight of a spider', she should be attacked with 'the shafts of ridicule'.

After passing a short and non-committal section on Religion, and judicious discussions of Address and Conversation, we reach the most influential parts of the book, on Exercise and on Air.

> The acquirements of literature, and of many arts, make the

lives of young people too sedentary; which impairs their strength, makes their countenances pale and bloated, and lays the foundation of many diseases; hence some hours should every day be appropriated to bodily exercises, and to relaxation of mind.

Such as tend to produce activity, and to promote the growth of the person in respect to height, are preferred in the schools for young ladies to those which render the system more robust and muscular.

Of these playing at ball, at shuttlecock, swinging as they sit on a cord or cushion, and dancing, in the open air in summer, and within doors in winter, are to be preferred. . . .

There are other modes of exertion which, though graceful in themselves, are not allowed to ladies by the fashion of this age and country; as skating on the ice in winter, swimming in summer, funambulation, or dancing on the straight rope. But walking with a stately measured step occasionally, like the march of soldiers, and reading aloud frequently rather in a theatric manner, as well as dancing and singing, will much contribute to give strength and grace to the muscles of locomotion, and of vocality. . . .

The strength and activity of young people not only depends on the perpetual exercise of their limbs, as described in the preceding sections, but on the purity of the air which they breathe. (pp. 68–9; 70; 70)

Darwin emphasizes the need for 'proper ventilation of the school-room, dining-room, bedrooms, and their other apartments'.

For this purpose it is convenient to saw off about one inch from the top of every door of these crowded rooms, and opposite to this aperture to nail along the top of the door a tin plate about two inches wide, rising at an angle of about 45 degrees; which will bend the current of air up towards the ceiling, where it will be mixed with the warm air of the room, and sink down amongst the society without the danger of giving cold to any one. And, besides these door-ventilators, the upper sashes of every window should always be let down a few inches, when the external weather will admit of it. (p. 72)

It was customary to keep bedrooms sealed, to exclude the

'noxious vapours' of the night; but Darwin wants to let in a breath of fresh air.

> The doors should be furnished with similar ventilators for the due admission of fresh air; and during the summer months a window should be kept a few inches open during the night as well as the day; the sash of which should have a bolt or other proper fastening for this purpose. Nor should the fireplace be stopped up at any season by a chimney-board, or a bag of straw; as many rooms are made to shut up so close that this is the only aperture by which fresh air is admitted. To this should be added, that the bed curtains should never be drawn close round the beds; which confine the air spoiled by frequent respiration, and the perspirable matter, like a noxious atmosphere over the unconscious sleepers. (pp. 72–3)

Darwin also saw the inefficiency of coal fires.

> As the quantity of air carried up a chimney is very great, owing to its being rendered so much lighter than the external atmosphere by the heat of the fire, strong currents of cold air press into the room at every chink of the doors and windows passing towards the fire. (pp. 73-4)

The remedy he favours is to narrow the chimneys just above the fire and thus reduce the upward flow of air.

The next section has a promising title, 'Care of the Shape', and begins well:

> Delicate young ladies are very liable to become awry at many boarding schools: this is occasioned principally by their being obliged too long to preserve an erect attitude, by sitting on forms for many hours together. To prevent this the school-seats should either have backs, on which they may occasionally rest themselves; or desks before them, on which they may occasionally lean. (pp. 75–6)

Darwin advises that their stays should be symmetrical, or preferably abolished altogether. He also recommends that children should not sleep on the same side each night. Growing awry is less of a problem today, but largely because Darwin's ideas on exercise have been adopted.

After caring for the shape, we come to covering for the shape. Dress should, Darwin thinks, be as graceful as fashion permits. And what is grace? He has an answer:

> As beauty consists of lines flowing in easy curves according to the analysis of Hogarth, those parts of dress which are composed of such lines are always agreeable. Thus a sash descending from one shoulder to the opposite hip, or a Grecian veil thrown back and winding carelessly down behind, are always beautiful; but a few white ostrich feathers rising on the head before, and a train of silk sweeping on the ground behind, add so much grace to a moving female figure as to attract all eyes with unceasing admiration. (p. 83)

Hard on the heels of this flowing train come Amusements. Darwin is rather scornful about playing cards, but very keen on chess:

> It has the experience of the remotest antiquity to recommend it, occasions no depraved passions, as it is not played for money; and by the caution perpetually necessary to watch your adversary, and the judgment required to contrive, arrange and manage your own affairs, employs and strengthens every part of the understanding. (p. 86)

When youthful play takes more destructive forms and rebuke is needed, Darwin is against harsh punishment. He thinks refractory children should be governed by the teacher's 'superiority of mind'.

Darwin is very knowledgeable about childish ailments. He offers a detailed treatment for lisping and squinting, and emphasizes the mental origin of stammering:

> That this impediment of pronunciation is altogether a disease of the mind, and not of the organs of speech, is shown by the stammerer being able to speak all words with perfect facility when alone, as in repeating a play; but begins to hesitate if anyone approaches, or even if he imagines that he is listened to. Those words also are most difficult to him to pronounce which he is conscious he cannot change for others, as when he is asked his own name, or the names of other persons, or of places; and the more so if he is aware that the hearer is impatient to be informed, and that he cannot conjecture the name before it is spoken. (pp. 96–7)

His explanations for other habits are more convincing and

practical than some of the far-fetched theories of later psychologists:

> By confinement in a school-room for many successive hours, and that without being suffered to vary their posture, some of the more active and lively children are liable to gain tricks of involuntary actions, as twitchings of the face, restless gesticulations of the limbs, biting their nails, etc., which are generally at first occasioned by the want of sufficient bodily exercise to expend the superfluous animal power, like the jumping of a squirrel in a cage; but are also liable to be caught by imitation of each other.
>
> To prevent this kind of deformity, children should be suffered to change their attitudes and situations more frequently; or to walk about as they get their lessons. To counteract it the earliest attention is necessary; as a few weeks frequently establish a bad habit which cannot be removed without great difficulty. (pp. 100–1)

Even the most fidgety child sleeps, however, so there is a long section about beds, with the accent on comfort. Darwin condemns parents who give children hard beds or set them alongside a cold outer wall.

The next item on the menu is diet, and the man who sobered the county of Derby offers one surprising recipe:

> For the drink of the more robust children water is preferable, and for the weaker ones, small beer. (p. 110)

This lapse apart, his ideas foreshadow modern practice. For example,

> Ripe fruits, or fruit pies, are peculiarly serviceable to the constitutions of children, as well as agreeable to their palates. . . . These should therefore be allowed to children at all seasons. (p. 111)

When Darwin discusses Economy, he covers not only 'prudent expenditure of their money', but also 'a parsimonious attention to the lapse of time': the time some young ladies spend in dressing is 'perfectly ridiculous'.

Darwin believes that school-education is better than being taught at home, largely because children learn from each other. He concludes his section on School Education with sentiments worthy of the plummiest Speech-Day orator:

> The Philosopher who despising the goods of fortune said, 'he was rich, though he carried about with him everything which he possessed,' meant to assert that strength of mind joined with strength of body were superior to any other advantages of life. A good education furnishes us with this inestimable treasure; it accompanies us at home, travels with us abroad; delights us in solitude, graces us in society; comforts us in misfortune, guards us in prosperity; contributes to the happiness of others, and ensures our own. (pp. 117–18)

This would make a good conclusion, but Darwin adds a catalogue of useful books, some of them recommended 'by ladies whose opinions I had reason to regard'. The list is not very interesting, except as a guide to the taste of the 1790s.

Darwin's *Female Education* is notable for its rich common sense, its many suggestions for reforms in the curriculum which are now completely accepted, such as the inclusion of science, modern languages and outdoor exercise, and some not yet accepted in Britain, such as a secular morality. Also, by his resolute opposition to the idea of the helpless and feather-brained female, Darwin was fanning the feebly-burning flame of women's emancipation: strangely enough he was destined to endure its full blaze, for Breadsall Church, where he was buried, was burnt to the ground by suffragettes in 1914.

5 Medicine Man

A Doctour of Phisik . . .
He was a very parfit practisour.
CHAUCER; *Canterbury Tales*

DARWIN was recognized as the leading medical practitioner of his day, and into his massive treatise *Zoonomia* (the first edition weighs 10 lb.) he distils a lifetime's rich experience in practical medicine. He devoted more of his physical and intellectual energy to this subject than to any other, and gained an almost legendary reputation. But some of his remedies for serious physical disease are desperate expedients, devised in face of a patient's almost certain death: he relies far too much on venesection (bleeding), purgatives (cathartics) and emetics. Still, some of his cures, even if they seem murderous to us, must have been effective or he would never have gained the reputation he did.

The picture is much brighter when we come to mental and psychosomatic illness. Darwin was most perceptive in his humane treatment of lunatics, who were at this time often locked up and beaten, in the hope of driving out the devils possessing them. Darwin noticed and treated neuroses which many modern doctors would fail to recognize at all, and he was also ahead of the average general practitioner today in appreciating how the mind can influence bodily disease.

Zoonomia was first published in two volumes. Volume I (1794) runs to 586 quarto pages, and Volume II (1796) is even bulkier, with 772 pages. The book enjoyed great success, with a second edition in 1796 and a third edition, in four octavo volumes totalling over 2,100 pages, in 1801. *Zoonomia* was 'translated into German, French and Italian, and was honoured by the Pope by being placed in the *Index Expurgatorius*',[21] as Charles Darwin wittily remarked. Another sign of the book's fame is the appearance of Thomas Brown's 560-page critical commentary, *Observations on the Zoonomia of Erasmus Darwin* (1798). Probably no other scientific book has evoked so solid a riposte so promptly.

Zoonomia is also notable for apparently being the first book with the decimal system of paragraphing, now so widely used in scientific textbooks and reports, e.g. 'section 4.2.3'.

In the preface Darwin justifies the book and apologizes for its faults:

> The purport of the following pages is an endeavour to reduce the facts belonging to ANIMAL LIFE into classes, orders, genera, and species; and, by comparing them with each other, to unravel the theory of diseases. . . .
>
> The want of a theory . . . to conduct the practice of medicine is lamented by its professors; for, as a great number of unconnected facts are difficult to be acquired, and to be reasoned from, the art of medicine is in many instances less efficacious under the direction of its wisest practitioners; and by that busy crowd, who either boldly wade in darkness, or are led into endless error by the glare of false theory, it is daily practised to the destruction of thousands; add to this the unceasing injury which accrues to the public by the perpetual advertisements of pretended nostrums; the minds of the indolent become superstitiously fearful of diseases which they do not labour under; and thus become the daily prey of some crafty empyric. . . .
>
> There are some modern practitioners who declaim against medical theory in general, not considering that to think is to theorize; and that no one can direct a method of cure to a person labouring under disease without thinking, that is, without theorizing; and happy therefore is the patient whose physician possesses the best theory. . . .
>
> A great part of this work has lain by the writer above twenty years, as some of his friends can testify: he had hoped by frequent revision to have made it more worthy the acceptance of the public; this however his other perpetual occupations have in part prevented, and may continue to prevent, as long as he may be capable of revising it; he therefore begs of the candid reader to accept of it in its present state, and to excuse any inaccuracies of expression, or of conclusion, into which the intricacy of his subject, the general imperfection of language, or the frailty he has in common with other men, may have betrayed him; and from which he has not the vanity to believe this treatise to be exempt. (I vii; viii; ix; ix–x)*

The least satisfactory feature of *Zoonomia* is Darwin's division of all bodily actions into four classes, which he calls Irritation, Sensation, Volition and Association. His best definition of the classes is as follows:

> The sensorium possesses four distinct powers, or faculties, which are occasionally exerted, and produce all the motions of the fibrous parts of the body; these are the faculties of producing fibrous motions in consequence of irritation, which is excited by external bodies; in consequence of sensation, which is excited by pleasure or pain; in consequence of volition, which is excited by desire or aversion; and in consequence of association, which is excited by other fibrous motions. We are hence supplied with four natural classes of diseases derived from their proximate causes; which we shall term those of irritation, those of sensation, those of volition, and those of association. (III v–vi)

This may not be clear to us, but Darwin seemed to have little trouble in popping each disease into its proper Class.

Volumes I and II in the third edition of *Zoonomia* (corresponding to Volume I of the first edition) contain forty discursive chapters on various medical or semi-medical topics. Sleep, reverie, vertigo and drunkenness receive a chapter each, along with more obvious subjects like the circulatory system, the stomach and intestines, and the liver.

The first subject discussed is a great favourite of his, ocular spectra.

> Place a circular piece of red silk about an inch in diameter on a sheet of white paper in a strong light. . . . Look for a minute on this area, or till the eye becomes somewhat fatigued, and then, gently closing your eyes, and shading them with your hand, a circular green area of the same apparent diameter becomes visible in the closed eye. This green area is the colour reverse to the red area, which had been previously inspected. . . . Hence it appears that a part of the retina, which had been fatigued by contraction in one direction, relieves itself by exerting the antagonist fibres, and producing a contraction in an

---

* Volume and page numbers after quotations refer to the third and most complete edition of *Zoonomia*.

opposite direction, as is common in the exertions of our muscles. (I 18)

From these simple beginnings he goes on to discuss the after-images of more complex shapes and forms, with colour diagrams. He also reprints a paper on ocular spectra from the *Philosophical Transactions* of the Royal Society, which records a long series of experiments on the subject. This paper was printed under the name of his son Dr Robert W. Darwin, and won him his election to the Royal Society, but Erasmus himself is believed to have had an important hand in the paper. These pioneering researches of the Darwins, which were widely praised, were extended by Goethe and Helmholtz.[22]

The after-effects of alcohol are specified just as precisely in the chapter on Drunkenness, one of the many diseases of irritation.

> The diseases in consequence of frequent inebriety, or of daily taking much vinous spirit without inebriety, consist in the paralysis which is liable to succeed violent stimulation. Organs whose actions are associated with others are frequently more affected than the organ which is stimulated into too violent action. . . . Hence in drunken people it generally happens that the secretory vessels of the liver become first paralytic, and a torpor with consequent gall-stones or schirrus of this viscus is induced with concomitant jaundice; otherwise it becomes inflamed in consequence of previous torpor, and this inflammation is frequently transferred to a more sensible part, which is associated with it, and produces the gout, or the rosy eruption of the face, or some other leprous eruption on the head, or arms, or legs. Sometimes the stomach is first affected, and paralysis of the lacteal system is induced: whence a total abhorrence from flesh-food, and general emaciation. In others, the lymphatic system is affected with paralysis, and dropsy is the consequence. In some inebriates the torpor of the liver produces pain without apparent schirrus, or gall-stones, or inflammation, or consequent gout, and in these epilepsy or insanity are often the consequence. . . . Mark what happens to a man who drinks a quart of wine or of ale, if he has not been habituated to it. He loses the use both of his limbs and of his understanding! He becomes a temporary idiot, and has a temporary stroke of the

FIG. 1. Erasmus Darwin at the age of thirty-eight, from the portrait by Joseph Wright of Derby, now in Darwin College, Cambridge. Another version of this painting is in the National Portrait Gallery.

FIG. 2. Mrs Elizabeth Pole, who became Erasmus Darwin's second wife in 1781, when she was thirty-four. From a portrait at Down House, Kent.

FIG. 3 (*left*). The silver medallion commemorating the bicentenary of the Lunar Society, struck by The Mint, Birmingham, in 1966. The medallion, two inches in diameter, carries the names of the fourteen authenticated 'Lunaticks'.

FIG. 4 (*below*). Silhouette of Dr Erasmus Darwin (*right*) playing chess with his second son, Erasmus. Chess was one of the doctor's favourite pastimes.

> palsy! And though he slowly recovers after some hours, is it not reasonable to conclude that a perpetual repetition of so powerful a poison must at length permanently affect him? If a person accidentally becomes intoxicated by eating a few mushrooms of a peculiar kind, a general alarm is excited, and he is said to be poisoned, and emetics are exhibited; but so familiarized are we to the intoxication from vinous spirit that it occasions laughter rather than alarm. (I 366–7; 368)

He believed that all liver disease was caused by alcohol and favoured an alcoholic interpretation of the myth of Prometheus:

> I shall conclude this section on the diseases of the liver induced by spirituous liquors, with the well-known story of Prometheus, which seems indeed to have been invented by physicians in those ancient times, when all things were clothed in hieroglyphic, or in fable. Prometheus was painted as stealing fire from heaven, which might well represent the inflammable spirit produced by fermentation; which may be said to animate or enliven the man of clay; whence the conquests of Bacchus, as well as the temporary mirth and noise of his devotees. But the after punishment of those who steal this accursed fire is a vulture gnawing the liver; and well allegorizes the poor inebriate lingering for years under painful hepatic diseases. (II 9)

If any drunkard was sufficiently appalled by these tirades to wish for a cure, Darwin could oblige:

> In the chronical debility brought on by drinking spirituous or fermented liquors, there is another golden rule by which I have successfully directed the quantity of spirit which they may safely lessen, for there is no other means by which they can recover their health. It should be premised that where the power of digestion in these patients is totally destroyed, there is not much reason to expect a return to healthful vigour.
>
> I have directed several of these patients to omit one fourth part of the quantity of vinous spirit they have been lately accustomed to, and if in a fortnight their appetite increases, they are advised to omit another fourth part; but if they perceive that their digestion becomes impaired from the want of

> this quantity of spirituous potation, they are advised to continue as they are, and rather bear the ills they have, than risk the encounter of greater. At the same time flesh-meat with or without spice is recommended, with Peruvian bark and steel in small quantities between their meals, and half a grain of opium, or a grain, with five or eight grains of rhubarb at night. (I 133)

This remedy is based on the principle that when a stimulus is repeated, its effect gradually wears off:

> But if a second dose of opium be exhibited before the fibres have regained their natural quantity of sensorial power, its effect will be much less than the former, because the spirit of animation or sensorial power is in part exhausted by the previous excess of exertion. Hence all medicines repeated too frequently gradually lose their effect, as opium and wine. Many things of disagreeable taste at first cease to be disagreeable by frequent repetition, as tobacco; grief and pain gradually diminish, and at length cease altogether, and hence life itself becomes tolerable. (I 106)

Both the sober and the tipsy reader might find much of interest in the chapter on instinct, which abounds with amusing stories about insects, like this waspish anecdote:

> One circumstance I shall relate which fell under my own eye, and showed the power of reason in a wasp, as it is exercised among men. A wasp, on a gravel walk, had caught a fly nearly as large as himself; kneeling on the ground I observed him separate the tail and the head from the body part, to which the wings were attached. He then took the body part in his paws, and rose about two feet from the ground with it; but a gentle breeze wafting the wings of the fly turned him round in the air, and he settled again with his prey upon the gravel. I then distinctly observed him cut off with his mouth, first one of the wings, and then the other, after which he flew away with it unmolested by the wind.
>
> Go, thou sluggard, learn arts and industry from the bee, and from the ant!
>
> Go, proud reasoner, and call the worm thy sister! (I 263)

So far the excerpts have been only mildly medical. But readers with strong stomachs can enjoy many messy case-histories, like this one:

> Mr. —— ate two putrid pigeons out of a cold pigeon-pie, and drank about a pint of beer and ale along with them, and immediately rode about five miles. He was then seized with vomiting, which was after a few periods succeeded by purging; these continued alternately for two hours; and the purging continued by intervals for six or eight hours longer. During this time he could not force himself to drink more than one pint in the whole; this great inability to drink was owing to the nausea, or inverted motions of the stomach, which the voluntary exertion of swallowing could seldom and with difficulty overcome; yet he discharged in the whole at least six quarts; whence came this quantity of liquid? First, the contents of the stomach were emitted, then of the duodenum, gall-bladder, and pancreas, by vomiting. After this the contents of the lower bowels; then the chyle that was in the lacteal vessels and in the receptacle of chyle, was regurgitated into the intestines by a retrograde motion of these vessels. And afterwards the mucus deposited in the cellular membrane, and on the surface of all the other membranes, seems to have been absorbed. . . . (I 414)

Many of the cases serve only to show the ineffectiveness of the medicines, but sometimes Darwin produces a logical and effective treatment:

> Mr. B—— had a haemorrhage from his kidney, and parted with not less than a pint of blood a day (by conjecture) along with his urine for above a fortnight: venesections, mucilages, balsams, preparations of lead, the bark*, alum, and dragon's blood, opiates, with a large blister on his loins, were separately tried, in large doses, to no purpose. He was then directed to bathe in a cold spring up to the middle of his body only, the upper part being covered, and the haemorrhage diminished at the first, and ceased at the second immersion.
>
> In this case the external capillaries were rendered quiescent by the coldness of the water, and thence a less quantity of blood was circulated through them; and the internal capillaries, or other glands, became quiescent from their irritative associations with the external ones; and the haemorrhage was stopped a sufficient time for the ruptured vessels to contract their apertures, or for the blood in those apertures to coagulate.

* Peruvian bark, quinine.

Just as good is his electrical treatment of Mr. S.:

> Mr. S., a gentleman between 40 and 50 years of age, had had the jaundice about six weeks, without pain, sickness, or fever; and had taken emetics, cathartics, mercurials, bitters, chalybeates, essential oil, and ether, without apparent advantage. On a supposition that the obstruction of the bile might be owing to the paralysis or torpid action of the common bile-duct, and the stimulants taken into the stomach seeming to have no effect, I directed half a score smart electric shocks from a coated bottle, which held about a quart, to be passed through the liver, and along the course of the common gall-duct, as near as could be guessed, and on that very day the stools became yellow; he continued the electric shocks a few days more, and his skin gradually became clear. (II 2–3)

Darwin was stimulating muscles with electricity twenty years before Galvani's experiments brought the subject into prominence in the 1790s. For an example of Darwin at work we can turn to a letter written by Wedgwood in 1779, when his baby daughter Mary Anne became partly paralysed. 'Dr. Darwin was here on Friday with my father and ordered our little girl to be electrified two or three times a day on the side affected and to be continued for some weeks. We are willing to flatter ourselves that she has received some benefit from electricity already, as she begins to move her arm and leg a little. . . .'[23]

Darwin was also ahead of the professionals in appreciating oxygen. In the 1790s most British chemists still clung to the phlogiston theory of air; they believed that a burning substance gives out 'phlogiston', and they rejected the true explanation, that it combines with oxygen. Even the great Priestley and that 'mighty chemist' Keir were 'phlogistians', and would not accept the new picture being offered by Lavoisier and the French school, as shown by the letters quoted in Chapter 2. Darwin, after paying lip-service to both sides, cuts through the fog of conflicting beliefs and goes on to give his own excellent description of the oxygenation of blood:

> First, by the late discoveries of Dr. Priestley, M. Lavoisier, and other philosophers, it appears that the basis of atmospheri-

> cal air, called oxygene, is received by the blood through the membranes of the lungs; and that by this addition the colour of the blood is changed from a dark to a light red. Secondly, that water possesses oxygene also as a part of its composition, and contains air likewise in its pores; when the blood of fish receives oxygene from the water, or from the air it contains, by means of their gills, in the same manner as the blood is oxygenated in the lungs of air-breathing animals; it changes its colour at the same time from a dark to a light red in the vessels of their gills, which constitute a pulmonary organ adapted to the medium in which they live. Thirdly, that the placenta consists of arteries carrying the blood to its extremities, and a vein bringing it back, resembling exactly in structure the lungs and gills above mentioned; and that the blood changes its colour from a dark to a light red in passing through these vessels.
>
> This analogy between the lungs and gills of animals, and the placenta of the fetus, extends through a great variety of other circumstances; thus air-breathing creatures and fish can live but a few minutes without air or water; or when they are confined in such air or water as has been spoiled by their own respiration; the same happens to the fetus, which, as soon as the placenta is separated from the uterus, must either expand its lungs, and receive air, or die. Hence from the structure, as well as the use of the placenta, it appears to be a respiratory organ, like the gills of fish, by which the blood in the fetus becomes oxygenated. (II 184–6)

Volumes III and IV in the third edition of *Zoonomia* (corresponding to Volume II of the 1796 edition) provide a detailed catalogue of diseases; and after each disease comes the treatment, plausible or otherwise, under the heading M.M., which stands for Materia Medica. The diseases are sifted into Classes, Orders and Genera, as in the Linnaean division of plants; but Darwin has not become the Linnaeus of medicine because his categories are fallacious. There are four Classes, one for each of the four faculties of the sensorium. The following examples are typical.

CLASS I: Diseases of Irritation

ORDO I: Increased irritation

GENUS I: With increased actions of the sanguiferous system

SPECIES

1.	Febris irritativa	Irritative fever
2.	Ebrietas	Drunkenness
3.	Haemorrhagia arteriosa	Arterial haemorrhage
4.	Haemoptoe arteriosa	Spitting of arterial blood
5.	Haemorrhagia narium	Bleeding from the nose. . . .

(III 4)

CLASS III: Diseases of Volition

ORDO I: Increased volition

GENUS II: With increased actions of the organs of sense

SPECIES

1.	Mania mutabilis	Mutable madness
2.	Studium inane	Reverie
3.	Vigilia	Watchfulness
4.	Erotomania	Sentimental love
5.	Amor sui	Vanity
6.	Nostalgia	Desire of home
7.	Spes religiosa	Superstitious hope
8.	Superbia stemmatis	Pride of family
9.	Ambitio	Ambition
10.	Maeror	Grief
11.	Taedium vitae	Irksomeness of life
12.	Desiderium pulchritudinis	Loss of beauty
13.	Paupertatis timor	Fear of poverty. . . .

(IV 3)

—plus twelve other species, just as ingenious. Certainly there was no fear of poverty of invention while Darwin was on the scene.

Darwin's remedies for minor diseases are often as good as those offered today, and he is up-to-date on childbirth too:

> *Parturitio*. Parturition is not a disease, it is a natural process, but is more frequently unfortunate in high life than amongst the middle class of females; which may be owing partly to fear with which the priests of Lucina* are liable to inspire the ladies of

* Roman goddess of childbirth.

fashion to induce them to lie-in in town; and partly to the bad air of London, to which they purposely resort.

There are however other causes, which render parturition more dangerous to the ladies of high life; such as their greater general debility from neglect of energetic exercise, their inexperience of the variations of cold and heat, and their seclusion from fresh air. To which must be added, that great source of the destruction of female grace and beauty, as well as of female health, the tight stays, and other bandages, with which they are generally tortured in their early years by the active folly of their friends, which by displacing many of the viscera impedes their actions, and by compressing them together produces adhesions of one part to another, and affects even the form and aperture of the bones of the pelvis, through which the nascent child must be protruded.

As parturition is a natural, not a morbid process, no medicine should be given, where there is no appearance of disease. The absurd custom of giving a powerful opiate without indication to all women, as soon as they are delivered is, I make no doubt, frequently attended with injurious, and sometimes with fatal consequences. (III 301–2)

For an example of a familiar disease, we can turn to measles:

Rubeola irritata, morbilli. The measles commence with sneezing, red eyes, dry hoarse cough, and are attended with sensitive irritated fever. On the fourth day, or a little later, small thick eruptions appear, scarcely eminent above the skin, and, after three days, changing into very small branny scales. . . . (III 392)

Darwin soon pulls out his lancet to begin the blood-letting:

The measles are usually attended with inflammatory fever with strong pulse, and bear the lancet in every stage of the disease. In the early periods of it, venesection renders the fever and cough less; and, if any symptoms of peripneumony occur, is repeatedly necessary; and at the decline of the disease, if a cough be left after the eruption has ceased, and the subsequent branny scales are falling off, venesection should be immediately used; which prevents the danger of consumption. At this time also change of air is of material consequence, and often

> removes the cough like a charm. . . . It is probable that inoculation might disarm the measles as much as the small-pox. . . . Some attempts have been made, but a difficulty seems to arise in giving the disease; the blood, I conjecture, would not infect, nor the tears; perhaps the mucous discharge from the nostrils might succeed; or a drop of warm water put on the eruptions, and scraped off again with the edge of a lancet; or if the branny scales were collected, and moistened with a little warm water? Further experiments on this subject would be worthy the public attention. (III 393–4)

We can applaud the last sentence, even if the frequent venesection makes our blood run cold.

Soon after measles comes scarlet fever:

> *Scarlatina maligna*. The malignant scarlet fever begins with inflamed tonsils; which are succeeded by dark drab-coloured sloughs from three to five lines in diameter, flat, or beneath the surrounding surface; and which conceal beneath them spreading gangrenous ulcers. The swellings of the tonsils are sensible to the eye and touch externally, and have an elastic rather than an oedematous feel, like parts in the vicinity of gangrenes. The pulse is very quick and weak, with delirium, and the patient generally dies in a few days; or if he recovers, it is by slow degrees, and attended with anasarca.
>
> M.M. A vomit once. Wine. Beer. Cyder. Opium. Bark, in small repeated doses. Small successive blisters, if the extremities are cooler than natural. Cool air on the hot parts of the skin, the cool extremities being at the same time covered. Iced lemonade. Broth. Custards. Milk. Jellies. Bread pudding. Chicken. . . . (III 397)

This is one of the many occasions when M.M. could stand for Memento Mori, rather than Materia Medica. The list of M.M. given above, daunting though it may seem, is merely an aperitif, for Darwin has two pages more to offer. However, he can scarcely be blamed for trying everything against such a desperate disease.

For a whiff of Darwin's famous experimental methods, we can turn to his powder generator for pulmonary consumption.

> I wished to try the external application of such powders to

> ulcers in the lungs, and constructed a box with a circulating brush in it, as described in the annexed plate; into this box two ounces of fine powder of Peruvian bark were put, and two drachms of cerussa in fine powder; on whirling the central brush, part of this was raised into a cloud of powder, and the patient, applying his mouth to one of the tin pipes rising out of the box, inhaled this powder twice a day into his lungs. I observed it did not produce any cough or uneasiness. This patient was in the last stage of consumption, and was soon tired of the experiment, nor have I had such patients as I wished for the repetition of it. Perhaps a fine powder of manganese, or of the flowers of zinc, or of lapis calaminaris, might be thus applied to ulcers of the lungs with greater advantage? (III 468–9)

Darwin has a diagram showing details of this powder puffer, and his patients, unless they were beyond cure, must have enjoyed at least a psychological boost when they saw that he had taken the trouble to construct a new gadget just for their complaint.

So far we have sampled only the diseases of Irritation and Sensation.

> We now step forward to consider the diseases of volition, that superior faculty of the sensorium, which gives us the power of reason, and by its facility of action distinguishes mankind from brute animals; which has effected all that is great in the world, and superimposed the works of art on the situations of nature. (IV 6)

Darwin is at his best on diseases of volition, for he recognizes that bodily ailments are often of mental origin. Many of his examples can still be read with profit or amusement, and need no comment:

> *Clamor*. Screaming from pain. The talkative animals, as dogs, and swine, and children, scream most, when they are in pain, and even from fear; as they have used this kind of exertion from their birth most frequently and most forcibly; and can therefore sooner exhaust the accumulation of sensorial power in the affected muscular or sensual organs by this mode of exertion....
>
> *Risus*. The pleasurable sensations, which occasion laughter, are perpetually passing into the bounds of pain; for pleasure and

pain are often produced by different degrees of the same stimulus; as warmth, light, aromatic or volatile odours, become painful by their excess; and the tickling on the soles of the feet in children is a painful sensation at the very time it produces laughter. When the pleasurable ideas, which excite us to laugh, pass into pain, we use some exertion, as a scream, to relieve the pain, but soon stop it again, as we are unwilling to lose the pleasure; and thus we repeatedly begin to scream, and stop again alternately. So that in laughing there are three stages, first of pleasure, then pain, then an exertion to relieve that pain. (IV 13–14)

Spes religiosa. Superstitious hope. This maniacal hallucination in its milder state produces, like sentimental love, an agreeable reverie; but when joined with works of supererogation, it has occasioned many enormities. In India devotees consign themselves by vows to most painful and unceasing tortures, such as holding up their hands, till they cannot retract them; hanging up by hooks put into the thick skin over their shoulders, sitting upon sharp points, and other self-torments. While in our part of the globe fasting and mortification, as flagellation, has been believed to please a merciful deity! The serenity with which many have suffered cruel martyrdoms is to be ascribed to this powerful reverie.

Mr. ——, a clergyman, formerly of this neighbourhood, began to bruise and wound himself for the sake of religious mortification, and passed much time in prayer, and continued whole nights alone in the church. As he had a wife and family of small children, I believed the case to be incurable; as otherwise the affection and employment in his family connections would have opposed the beginning of this insanity. He was taken to a madhouse without effect, and after he returned home, continued to beat and bruise himself, and by this kind of mortification, and by sometimes long fasting, he at length became emaciated and died. I once told him in conversation, that 'God was a merciful being, and could not delight in cruelty, but that I supposed he worshipped the devil.' He was struck with this idea, and promised me not to beat himself for three days, and I believe kept his word for one day. If this idea had been frequently forced on his mind, it might probably have been of service.

When these works of supererogation have been of a public

nature, what cruelties, murders, massacres, has not this insanity introduced into the world!—A commander, who had been very active in leading and encouraging the bloody deeds of St. Bartholomew's day at Paris, on confessing his sins to a worthy ecclesiastic on his death-bed, was asked, 'Have you nothing to say about St. Bartholomew?' 'On that day,' he replied, 'God Almighty was obliged to me!'—The fear of Hell is another insanity, which will be spoken of below. (IV 83–4)

Orci timor. The fear of Hell. Many theatric preachers among the Methodists successfully inspire this terror, and live comfortably upon the folly of their hearers. In this kind of madness the poor patients frequently commit suicide; although they believe they run headlong into the Hell which they dread! Such is the power of oratory, and such the debility of the human understanding!

Those who suffer under this insanity are generally the most innocent and harmless people; who are then liable to accuse themselves of the greatest imaginary crimes, and have so much intellectual cowardice that they dare not reason about those things which they are directed by their priests to believe, however contradictory to human apprehension, or derogatory to the great Creator of all things. The maniacal hallucination at length becomes so painful that the poor insane flies from life to become free from it.

M.M. Where the intellectual cowardice is great, the voice of reason is ineffectual; but that of ridicule may save many from those mad-making doctors; though it is too weak to cure those who are already hallucinated. Foote's Farces* are recommended for this purpose. (IV 102–3)

Credulitas. Credulity. Life is short, opportunities of knowledge rare; our senses are fallacious, our reasonings uncertain; man therefore struggles with perpetual error from the cradle to the coffin. He is necessitated to correct experiment by analogy, and analogy by experiment; and not always to rest satisfied in the belief of facts even with this two-fold testimony, till future opportunities, or the observations of others, concur in their support.

* Samuel Foote (1720–77), sometimes called 'the English Aristophanes', was a prolific author of farces, including one against the Methodists.

Ignorance and credulity have ever been companions, and have misled and enslaved mankind; philosophy [i.e. science] has in all ages endeavoured to oppose their progress, and to loosen the shackles they had imposed; philosophers have on this account been called unbelievers: unbelievers of what? of the fictions of fancy, of witchcraft, hobgoblins, apparitions, vampires, fairies; of the influence of stars on human actions, miracles wrought by the bones of saints, the flights of ominous birds, the predictions from the bowels of dying animals, expounders of dreams, fortune-tellers, conjurors, modern prophets, necromancy, cheiromancy, animal magnetism, metallic tractors, with endless variety of folly? These they have disbelieved and despised, but have ever bowed their hoary heads to Truth and Nature. . . .

In regard to religious matters, there is an intellectual cowardice instilled into the minds of the people from their infancy; which prevents their inquiry: credulity is made an indispensable virtue; to inquire or exert their reason in religious matters is denounced as sinful; and in the catholic church is punished with more severe penances than moral crimes. . . .

M.M. The method of cure is to increase our knowledge of the laws of nature, and our habit of comparing whatever ideas are presented to us with those known laws, and thus to counteract the fallacies of our senses, to emancipate ourselves from the false impressions which we have imbibed in our infancy, and to set the faculty of reason above that of imagination. (IV 152; 155; 156)

Darwin's kind treatment of the mentally ill is shown by the tone of this comment on a case of hallucination:

Miss G—— . . . said, as I once sat by her, 'My head is fallen off, see it is rolled to that corner of the room, and the little black dog is nibbling the nose off'. On my walking to the place which she looked at, and returning, and assuring her that her nose was unhurt, she became pacified, though I was doubtful whether she attended to me. (IV 74)

Darwin is not always so mild, for if the mental condition provokes convulsions it receives the full frightening treatment.

M.M. To remove the peculiar pain which excites the convulsions. Venesection. An emetic. A cathartic with calomel. Warm-bath. Opium in large quantities, beginning with smaller ones. Mercurial frictions. Electricity. Cold-bath in the paroxysm; or cold aspersion. (IV 16–17)

Among Darwin's diseases is *taedium vitae*, which he recognizes as a frequent cause of suicide.

> *Taedium vitae*. Ennui. Irksomeness of life. The inanity of sublunary things has afforded a theme to philosophers, moralists, and divines, from the earliest records of antiquity; 'Vanity of vanities!' says the preacher, 'all is vanity!'. . . . Nor is there anyone, I suppose, who has passed the meridian of life, who has not at some moments felt the nihility of all things. (IV 90)

His cure is even more pertinent today, when children 'know it all' so early in life:

> M.M. Some restraint in exhausting the usual pleasures of the world early in life. The agreeable cares of a matrimonial life. The cultivation of science, as of chemistry, natural philosophy, natural history, which supplies an inexhaustible source of pleasurable novelty, and relieves ennui by the exertions it occasions. (IV 95–6; wording is that of second edition, Vol. II, p. 374)

'Loss of beauty' also features among the diseases 'painfully felt by ladies'. However, fashions change and one of his remedies, removing sun-tan with hydrochloric acid, will scarcely commend itself to the modern girl; even in the eighteenth century the remedy must have been as 'painfully felt' as the condition it combated.

> The tan of the skin occasioned by the sun may be removed by lemon juice evaporated by the fire to half its original quantity, or by diluted marine acid; which cleans the cuticle, by eroding its surface, but requires much caution in the application; the marine acid must be diluted with water, and then put upon the hand or face; after a second of time, as soon as the tan disappears, the part must be washed with a wet towel and much warm water. (IV 97)

Today, when skin colour is such a burning issue, Darwin's idea of bleaching is liable to inflame both the skin and the animosity of racialists.

From Volition we pass to Association, and one of the best sections is on gout (its association of course is with alcohol), because Darwin reports his own experience.

> As example has more forcible effect than simple assertion, I shall now concisely relate my own case. . . . E.D. was about forty years of age, when he was first seized with a fit of the gout. The ball of his right great toe was very painful, and much swelled and inflamed, which continued five or six days in spite of venesection, a brisk cathartic with ten grains of calomel, and the application of cold air and cold water to his foot. He then ceased to drink ale or wine alone; confining himself to small-beer, or wine diluted with about thrice its quantity of water. In about a year he suffered two other fits of the gout, in less violent degree. He then totally abstained from all fermented liquors, not even tasting small-beer, or a drop of any kind of wine; but ate plentifully of flesh-meat, and all kinds of vegetables, and fruit, using for his drink at meals chiefly water alone, or lemonade, or cream water; with tea and coffee between them as usual.
>
> By this abstinence from fermented liquors he kept quite free from the gout for fifteen or sixteen years; and then began to take small-beer mixed with water occasionally, or wine and water, or perry and water, or cyder and water; by which indulgence after a few months he had again a paroxysm of gout, which continued about three days in the ball of his toe; which occasioned him to return to his habit of drinking water, and he has now for above twenty years kept in perpetual health, except accidental colds from the changes of the seasons. Before he abstained from fermented or spirituous liquors, he was frequently subject to the piles, and to the gravel, neither of which he had since experienced. (IV 209–10)

Apart from a few such juicy specimens, however, the diseases of association are rather a feeble fraternity. The shakiest of all is *pallor et tremor a timore*, for which Darwin offers the pleasing treatment: 'Opium. Wine. Food. Joy.'

The section on diseases of association ends with a ninety-page supplement, entitled 'Sympathetic Theory of Fever', which had great influence in its time. Today it is open to the objection that Darwin seems to treat the symptoms rather than their cause. Two of his suggestions have subsequently proved of value, however. The first of these is the old idea of blood transfusion, which he expounds in some detail.

The second idea seems at first sight to be rather a giddy one:

> Another experiment I have frequently wished to try, which cannot be done in private practice, and which I therefore recommend to some hospital physician; and that is, to endeavour to still the violent actions of the heart and arteries, after due evacuations by venesection and cathartics, by gently compressing the brain. This might be done by suspending a bed, so as to whirl the patient round with his head most distant from the centre of motion, as if he lay across a mill-stone, as described in Sect. XVIII. 20. For this purpose a perpendicular shaft armed with iron gudgeons might have one end pass into the floor, and the other into a beam in the ceiling, with an horizontal arm, to which a small bed might be readily suspended.
>
> By thus whirling the patient with increasing velocity sleep might be produced, and probably the violence of the actions of the heart and arteries might be diminished in inflammatory fevers . . . the lives of thousands might probably be saved by thus extinguishing the exacerbations of febrile paroxysms, or preventing the returns of them. (IV 436–7)

The relevant passage in Section XVIII. 20 is as follows:

> Another way of procuring sleep mechanically was related to me by Mr. Brindley, the famous canal engineer, who was brought up to the business of a mill-wright; he told me that he had more than once seen the experiment of a man extending himself across the large stone of a corn-mill, and that by gradually letting the stone whirl, the man fell asleep before the stone had gained its full velocity, and he supposed would have died without pain by the continuance or increase of the motion. In this case the centrifugal motion of the head and feet must accumulate the blood in both those extremities of the body, and thus compress the brain. (I 314)

In fact Darwin's idea of whirling became quite popular, especially in the treatment of mental illness, and Watt made a detailed engineering design of a rotative couch to Darwin's specification, shown in Fig. 7. Now we have come full circle, and the centrifuge is again to be used therapeutically, to create artificial gravity in weightless space vehicles. In his history of the centrifuge in aerospace medicine, Dr W. J. White has given Darwin full credit: 'As far as can be ascertained, Darwin's analysis marked the beginning of the therapeutic use of centrifugation'.[24]

Zoonomia ends with 118 pages of the Materia Medica, defined by Darwin as 'all those substances which may contribute to the restoration of health'—with the emphasis on *may*. This medicine chest has seven compartments. The first holds the Nutrientia: these are mostly foods, but oxygen and blood transfusion also find their way in. Food leads Darwin to reflect upon property and the inequality of man:

> However the arts of cookery and of grinding may increase or facilitate the nourishment of mankind, the great source of it is from agriculture. In the savage state, where men live solely by hunting, I was informed by Dr. Franklin that there was seldom more than one family existed in a circle of five miles diameter; which in a state of pasturage would support some hundred people, and in a state of agriculture many thousands. . . .
>
> But pasturage cannot exist without property both in the soil and the herds which it nurtures; and for the invention of arts, and production of tools necessary to agriculture, some must think, and others labour; and as the efforts of some will be crowned with greater success than that of others, an inequality of the ranks of society must succeed; but this inequality of mankind in the present state of the world is too great for the purposes of producing the greatest quantity of human nourishment, and the greatest sum of human happiness; there should be no slavery at one end of the chain of society, and no despotism at the other. By the future improvements of human reason such governments may possibly hereafter be established, as may a hundred-fold increase the numbers of mankind, and a thousand-fold their happiness. (II 415; 416)

The second class of medicaments is the Incitantia, which 'increase the irritative motions': these include opium and alcohol, not to mention love and joy.

The third and fourth classes are the Secernentia and Sorbentia: these 'increase the irritative motions which constitute' secretion and absorption respectively. The lists of items extend to many pages, and range from ginger to hopeless distress (Secernentia) and from acid of vitriol to electric sparks (Sorbentia). The remaining classes, Invertantia, Revertantia and Torpentia, need scarcely detain us, but at least they show that Darwin had a large if ineffective stock of medicines.

I have extracted a few ounces of choice morsels from the ten-pound corpus of *Zoonomia*, and Darwin might not have approved of my surgical principles. However, *Zoonomia*, more than any other of his works, is suitable for severe cutting, and the extracts quoted here are quite enough to demonstrate its virtues and defects.

In *Zoonomia* Darwin gives his considered opinions on medical matters. Some uncensored reports of actual treatments are also available, if we pry into his manuscript Commonplace Book. The most murderous is this one, probably in 1776:

> Mrs. — was asthmatic and dropsical, but did not appear near her end. She took four draughts of the decoction of Foxglove. . . . She vomited two or three times, and then purged twice and died upon the close-stool. (p. 8)

In contrast, Mrs. Stubs, who had a bad sore throat, escaped lightly:

> She has the greatest number of rotten teeth in the highest degree of decay that I ever saw in one mouth. I have advised her teeth to be drawn. . . . (p. 85)

and we hope they took the advice. The next extract refers to a case of pleurisy:

> Mr. Jauncey was blooded nine times, he then became feeble, with obscure hiccup for a day and twitching of the corners of his mouth into a kind of risus sardonicus. . . . (p. 66)

The poor man didn't have much to laugh about: no wonder the smile was sardonic! However, he recovered.

These lurid excerpts are unfair to Darwin: in fact he was one of the kindest of doctors, outraged by needless cruelty and always ready to go to the help of a patient, even if it meant a journey of forty miles on Christmas Day:

> I have this day (Dec. 25, 1778) seen a Mrs. Riley of Stafford, who has a diseased digestion . . . . (p. 51)

6 Evolution

> There was no break in that long line of law
> Between the first life drifting in the sea,
> And man, proud man, the crowning form of earth.
>
> ALFRED NOYES; *The Torch-Bearers: Erasmus Darwin*

THE ancient Greeks had the idea that the species now seen on Earth evolved gradually from primitive forms. During the next two thousand years the idea was revived from time to time, but it made little headway in seventeenth and eighteenth century Europe, for two reasons. First, the Christian Church decreed that species were created by God and immutable. Second, the scientific classification of plant species by Linnaeus inevitably tended to emphasize the fixity of species.

Erasmus Darwin propounded a theory of evolution that was more complete than any earlier version and was, as far as it went, very close to the modern form of the theory—indeed he was nearer the modern view than either Lamarck (a few years later) or his own grandson Charles (some sixty years later).[25] Erasmus was less committed to belief in the inheritance of acquired characters and he correctly estimated the time scale of evolution as 'millions of ages'.

Erasmus had the right ideas, but failed to produce enough evidence to convince the world; his son Robert took over nearly all Erasmus's ideas (not only evolution, but also abstinence from alcohol and from Christianity); Robert's son Charles, imbued with evolutionary ideas from an early age by his father, developed those ideas and clothed them with a mass of evidence, winning himself an honoured and deserved place as one of the world's greatest scientists. The nineteenth-century admirers of Charles Darwin rightly emphasized that his evidence was much more extensive and convincing than that of Erasmus, and proceeded to lock grandfather's skeleton securely in the cupboard. Once Charles Darwin's greatness was fully accepted, his admirers could safely have allowed Erasmus to emerge and take his rightful place as a lineal predecessor, in theory as well as in reality. But censorship is habit-forming, and many of Charles Darwin's champions continued to suppress Erasmus long after the need had passed, with the result that he has been consistently under-rated as an evolutionist. A whole book could be written on the links between Erasmus's and Charles's theories of evolution;[26] but here we must leave Erasmus's words to speak for themselves.

His main pronouncements on evolution come in Chapter 39 of *Zoonomia*, entitled 'Of generation'. He begins badly by suggesting that the female's role in reproduction is merely to provide the nutritive matter for the embryo, while its characteristics are determined by the male, and may even be influenced by the imagination of the male at the time of conception. But he later abandons this false position, admitting that the 'nutriment' also has an important influence on the physical and mental characteristics of the offspring. In reading the following excerpts we should remember this illogicality, and also that Darwin uses the terms irritation, sensation, volition and association in the specialized senses described in the previous chapter.

> When we revolve in our minds, first, the great changes which we see naturally produced in animals after their nativity, as in the production of the butterfly with painted wings from the crawling caterpillar; or of the respiring frog from the subnatant tadpole; from the feminine boy to the bearded man, and from the infant girl to the lactescent woman; both which changes may be prevented by certain mutilations of the glands necessary to reproduction.
>
> Secondly, when we think over the great changes introduced into various animals by artificial or accidental cultivation, as in horses, which we have exercised for the different purposes of strength or swiftness, in carrying burthens or in running races; or in dogs, which have been cultivated for strength and courage, as the bull-dog; or for acuteness of his sense of smell, as the hound and spaniel; or for the swiftness of his foot, as the greyhound; or for his swimming in the water, or for drawing snow-sledges, as the rough-haired dogs of the north; or lastly, as a play-dog for children, as the lap-dog; with the changes of the forms of the cattle which have been domesticated from the

greatest antiquity, as camels, and sheep; which have undergone so total a transformation that we are now ignorant from what species of wild animals they had their origin. Add to these the great changes of shape and colour, which we daily see produced in smaller animals from our domestication of them, as rabbits, or pigeons; or from the difference of climates and even of seasons; thus the sheep of warm climates are covered with hair instead of wool; and the hares and partridges of the latitudes which are long buried in snow, become white during the winter months; add to these the various changes produced in the forms of mankind, by their early modes of exertion; or by the diseases occasioned by their habits of life; both of which became hereditary, and that through many generations. Those who labour at the anvil, the oar, or the loom, as well as those who carry sedan-chairs, or who have been educated to dance upon the rope, are distinguishable by the shape of their limbs; and the diseases occasioned by intoxication deform the countenance with leprous eruptions, or the body with tumid viscera, or the joints with knots and distortions.

Thirdly, when we enumerate the great changes produced in the species of animals before their nativity; these are such as resemble the form or colour of their parents, which have been altered by the cultivation or accidents above related, and are thus continued to their posterity. Or they are changes produced by the mixture of species as in mules; or changes produced probably by the exuberance of nourishment supplied to the fetus, as in monstrous births with additional limbs; many of these enormities of shape are propagated, and continued as a variety at least, if not as a new species of animal. I have seen a breed of cats with an additional claw on every foot; of poultry also with an additional claw, and with wings to their feet; and of others without rumps. Mr. Buffon mentions a breed of dogs without tails, which are common at Rome and at Naples, which he supposes to have been produced by a custom long established of cutting their tails close off. There are many kinds of pigeons, admired for their peculiarities, which are monsters thus produced and propagated. . . .

When we consider all these changes of animal form, and innumerable others which may be collected from the books of natural history; we cannot but be convinced that the fetus or embryon is formed by apposition of new parts, and not by the distention of a primordial nest of germes, included one within another, like the cups of a conjurer.*

Fourthly, when we revolve in our minds the great similarity of structure which obtains in all the warm-blooded animals, as well quadrupeds, birds, and amphibious animals, as in mankind; from the mouse and bat to the elephant and whale; one is led to conclude that they have alike been produced from a similar living filament. In some this filament in its advance to maturity has acquired hands and fingers, with a fine sense of touch, as in mankind. In others it has acquired claws or talons, as in tygers and eagles. In others, toes with an intervening web, or membrane, as in seals and geese. In others it has acquired cloven hoofs, as in cows and swine; and whole hoofs in others, as in the horse. While in the bird kind this original living filament has put forth wings instead of arms or legs, and feathers instead of hair. In some it has protruded horns on the forehead instead of teeth in the fore part of the upper jaw; in others tushes instead of horns; and in others beaks instead of either. And all this exactly as is daily seen in the transmutations of the tadpole, which acquires legs and lungs, when he wants them; and loses his tail, when it is no longer of service to him.

Fifthly, from their first rudiment, or primordium, to the termination of their lives, all animals undergo perpetual transformations; which are in part produced by their own exertions in consequence of their desires and aversions, of their pleasures and their pains, or of irritations, or of associations; and many of these acquired forms or propensities are transmitted to their posterity.

As air and water are supplied to animals in sufficient profusion, the three great objects of desire, which have changed the forms of many animals by their exertions to gratify them, are those of lust, hunger, and security. A great want of one part of the animal world has consisted in the desire of the exclusive possession of the females; and these have acquired weapons to combat each other for this purpose, as the very thick, shield-like, horny skin on the shoulder of the boar is a defence only against animals of his own species, who strike obliquely upwards, nor are his tushes for other purposes, except to defend

* This erroneous theory of 'pre-formation' was widely believed in Darwin's day.

himself, as he is not naturally a carnivorous animal. So the horns of the stag are sharp to offend his adversary, but are branched for the purpose of parrying or receiving the thrusts of horns similar to his own, and have therefore been formed for the purpose of combating other stags for the exclusive possession of the females; who are observed, like the ladies in the times of chivalry, to attend the car of the victor.

The birds which do not carry food to their young, and do not therefore marry, are armed with spurs for the purpose of fighting for the exclusive possession of the females, as cocks and quails. It is certain that these weapons are not provided for their defence against other adversaries, because the females of these species are without this armour. The final cause of this contest amongst the males seems to be, that the strongest and most active animal should propagate the species, which should thence become improved.

Another great want consists in the means of procuring food, which has diversified the forms of all species of animals. Thus the nose of the swine has become hard for the purpose of turning up the soil in search of insects and of roots. The trunk of the elephant is an elongation of the nose for the purpose of pulling down the branches of trees for his food, and for taking up water without bending his knees. Beasts of prey have acquired strong jaws or talons. Cattle have acquired a rough tongue and a rough palate to pull off the blades of grass, as cows and sheep. Some birds have acquired harder beaks to crack nuts, as the parrot. Others have acquired beaks adapted to break the harder seeds, as sparrows. Others for the softer seeds of flowers, or the buds of trees, as the finches. Other birds have acquired long beaks to penetrate the moister soils in search of insects or roots, as woodcocks; and others broad ones to filtrate the water of lakes, and to retain aquatic insects, as ducks. All which seem to have been gradually produced during many generations by the perpetual endeavour of the creatures to supply the want of food, and to have been delivered to their posterity with constant improvement of them for the purposes required.

The third great want amongst animals is that of security, which seems much to have diversified the forms of their bodies and the colour of them; these consist in the means of escaping other animals more powerful than themselves. Hence some animals have acquired wings instead of legs, as the smaller birds, for the purpose of escape. Others great length of fin, or of membrane, as the flying fish, and the bat. Others great swiftness of foot, as the hare. Others have acquired hard or armed shells, as the tortoise and the echinus marinus.

Mr. Osbeck, a pupil of Linnaeus, mentions the American frog-fish, lophius histrio, which inhabits the large floating islands of sea-weed about the Cape of Good Hope, and has fulcra resembling leaves, that the fishes of prey may mistake it for the sea-weed, which it inhabits. *Voyage to China*, p. 113.

The contrivances for the purposes of security extend even to vegetables, as is seen in the wonderful and various means of their concealing or defending their honey from insects,* and their seeds from birds. On the other hand swiftness of wing has been acquired by hawks and swallows to pursue their prey; and a proboscis of admirable structure has been acquired by the bee, the moth, and the humming bird, for the purpose of plundering the nectaries of flowers. All which seem to have been formed by the original living filament, excited into action by the necessities of the creatures which possess them, and on which their existence depends.

From thus meditating on the great similarity of the structure of the warm-blooded animals, and at the same time of the great changes they undergo both before and after their nativity; and by considering in how minute a portion of time many of the changes of animals above described have been produced; would it be too bold to imagine that in the great length of time since the earth began to exist, perhaps millions of ages before the commencement of the history of mankind, would it be too bold to imagine that all warm-blooded animals have arisen from one living filament, which THE GREAT FIRST CAUSE endued with animality, with the power of acquiring new parts, attended with new propensities, directed by irritations, sensations, volitions, and associations; and thus possessing the faculty of continuing to improve by its own inherent activity, and of delivering down those improvements by generation to its posterity, world without end? (*Zoonomia* II 233–40)

Having shown how species are improved by the success of the strongest animals, or those best adapted to their environment, Darwin goes on to discuss fish, insects, etc., in similar

* Darwin's ideas about honey were wrong. See Chapter 7.

terms, and finally vegetables. After quoting Linnaeus's remarks that vegetable species were changed and increased by their 'intermarriages', he continues:

> Many other changes seem to have arisen in them by their perpetual contest for light and air above ground, and for food or moisture beneath the soil. . . . From these one might be led to imagine that each plant at first consisted of a single bulb or flower to each root, as the gentianella and daisy; and that in the contest for air and light new buds grew on the old decaying flower stem, shooting down their elongated roots to the ground, and that in process of ages tall trees were thus formed, and an individual bulb became a swarm of vegetables. Other plants, which in this contest for light and air were too slender to rise by their own strength, learned by degrees to adhere to their neighbours, either by putting forth roots like the ivy, or by tendrils like the vine, or by spiral contortions like the honeysuckle; or by growing upon them like the mistletoe, and taking nourishment from their barks; or by only lodging or adhering on them, and deriving nourishment from the air, as tillandsia.
>
> Shall we then say that the vegetable living filament was originally different from that of each tribe of animals above described? And that the productive living filament of each of those tribes was different originally from the other? Or, as the earth and ocean were probably peopled with vegetable productions long before the existence of animals; and many families of these animals long before other families of them, shall we conjecture that one and the same kind of living filament is and has been the cause of all organic life?
>
> If this gradual production of the species and genera of animals be assented to, a contrary circumstance may be supposed to have occurred, namely, that some kinds by the great changes of the elements may have been destroyed. This idea is shown to our senses by contemplating the petrifactions of shells, and of vegetables, which may be said, like busts and medals, to record the history of remote times. Of the myriads of belemnites, cornua ammonis, and numerous other petrified shells, which are found in the masses of limestone which have been produced by them, none now are ever found in our seas, or in the seas of other parts of the world, according to the observations of many naturalists. Some of whom have imagined that most of the

> inhabitants of the sea and earth of very remote times are now extinct; as they scarcely admit that a single fossil shell bears a strict similitude to any recent ones, and that the vegetable impressions or petrifactions found in iron-ores, clay, or sandstone, of which there are many of the fern kind, are not similar to any plants of this country, nor accurately correspond with those of other climates, which is an argument countenancing the changes in the forms, both of animals and vegetables, during the progressive structure of the globe which we inhabit. (II 243–5)

In the third edition of *Zoonomia* (1801) Darwin produces a new theory of generation in terms of male and female 'fibrils' or 'molecules'. He has now abandoned the 'nutriment' and has in fact come close to the modern view, if his 'molecules' (which he intended to mean 'small particles') are read as DNA molecules.

> On considering the reproduction of vegetable buds and seeds, of some insects, and of more perfect animals, the modes of generation may be divided into solitary and sexual. . . .
>
> [In solitary] reproduction I suppose that fibrils with formative appetencies, and molecules with formative aptitudes or propensities, produced by, or detached from, various essential parts of their respective systems, float in the vegetable or insect blood. . . . (II 304)
>
> [In sexual reproduction] we may safely conclude . . . that as these fibrils or molecules floated in the circulating blood of the parents, they were collected separately by appropriated glands of the male or female; and that finally on their mixture in the matrix the new embryon was generated, resembling in some parts the form of the father, and in other parts the form of the mother, according to the quantity or activity of the fibrils or molecules at the time of their conjunction. (II 295)

Darwin emphasizes the importance of having two sexes to provide variety in the offspring:

> hence those hermaphrodite insects, if they could have produced young without a mother, would not have been capable of that change or improvement, which is seen in all other animals, and in those vegetables which are procreated by the male embryon received and nourished by the female. And it is hence probable,

> that if vegetables could only have been produced by buds and bulbs, and not by sexual generation, that there would not at this time have existed one thousandth part of their present number of species. (II 262–3)

This point is elaborated in *Phytologia* (see Chapter 7).

Darwin also gropes towards the statistical approach needed to establish the Mendelian theory of heredity:

> One might hence suspect that if Blumenbach had attended to the propagation of the polypus, which he had composed of two half polypi, that the young progeny might have possessed two colours resembling the compound parent, like the different caudexes of ingrafted trees; an experiment well worthy repeated observation. (II 282)

Darwin's evolutionary theory is most attractively presented in his last long poem, *The Temple of Nature*, which traces the progress of life from microscopic specks in primordial seas to its present state. Darwin remarks in the preface that his aim 'is simply to amuse', and he certainly achieves this aim in some of his more comic couplets. However, the poem is also a remarkably correct account of evolutionary development. The story begins with the formation of Earth:

> Ere Time began, from flaming chaos hurl'd
> Rose the bright spheres, which form the circling world;
> Earths from each sun with quick explosions burst,
> And second planets issued from the first.
> Then, whilst the sea at their coeval birth,
> Surge over surge, involv'd the shoreless earth;
> Nurs'd by warm sun-beams in primeval caves
> Organic Life began beneath the waves. . . .
> Hence without parent by spontaneous birth
> Rise the first specks of animated earth;
> From Nature's womb the plant or insect swims,
> And buds or breathes, with microscopic limbs.
>
> (Canto I, lines 227–34, 247–50)

The following extracts show the calm certainty of Darwin's belief in evolution from the most primitive forms to Man.

> Organic Life beneath the shoreless waves
> Was born and nurs'd in Ocean's pearly caves;

First forms minute, unseen by spheric glass,
Move on the mud, or pierce the watery mass;
These, as successive generations bloom,
New powers acquire, and larger limbs assume;
Whence countless groups of vegetation spring,
And breathing realms of fin, and feet, and wing.

Thus the tall Oak, the giant of the wood,
Which bears Britannia's thunders on the flood;
The Whale, unmeasured monster of the main,
The lordly Lion, monarch of the plain,
The Eagle soaring in the realms of air,
Whose eye undazzled drinks the solar glare,
Imperious man, who rules the bestial crowd,
Of language, reason, and reflection proud,
With brow erect, who scorns this earthy sod,
And styles himself the image of his God;
Arose from rudiments of form and sense,
An embryon point, or microscopic ens!

Now in vast shoals beneath the brineless tide,
On earth's firm crust testaceous tribes* reside;
Age after age expands the peopled plain,
The tenants perish, but their cells remain;
Whence coral walls and sparry hills ascend
From pole to pole, and round the line extend.

(I 296–320)

Then the host of tiny sea-creatures

Leaves the cold caverns of the deep, and creeps
On shelving shores, or climbs on rocky steeps.
As in dry air the sea-born stranger roves,
Each muscle quickens, and each sense improves;
Cold gills aquatic form respiring lungs,
And sounds aerial flow from slimy tongues.

(I 329–34)

One of the many notes usefully amplifies the argument:

* *Testaceous* means 'possessing shells'. Darwin is referring to the coral insects and the shells that go to make limestone.

After islands or continents were raised above the primeval ocean, great numbers of the most simple animals would attempt to seek food at the edges or shores of the new land, and might thence gradually become amphibious; as is now seen in the frog, who changes from an aquatic animal to an amphibious one; and in the gnat, which changes from a natant to a volant state.

At the same time new microscopic animalcules would immediately commence wherever there was warmth and moisture, and some organic matter, that might induce putridity.* Those situated on dry land, and immersed in dry air, may gradually acquire new powers to preserve their existence; and by innumerable successive reproductions for some thousands, or perhaps millions of ages, may at length have produced many of the vegetable and animal inhabitants which now people the earth.

(I 327, note)

In the second canto Darwin explains how sex brings in variety, emphasizing that asexual reproduction leads to repetition and hereditary diseases. This topic is discussed in more detail in *Phytologia* (Chapter 7).

The struggle for existence described in *Zoonomia* is more dramatically portrayed in Canto IV of *The Temple of Nature*:

The wolf, escorted by his milk-drawn dam,
Unknown to mercy, tears the guiltless lamb;
The towering eagle, darting from above,
Unfeeling rends the inoffensive dove;
The lamb and dove on living nature feed,
Crop the young herb, or crush the embryon seed.
Nor spares the loud owl in her dusky flight,
Smit with sweet notes, the minstrel of the night;
Nor spares, enamour'd of his radiant form,
The hungry nightingale the glowing worm;
Who with bright lamp alarms the midnight hour,
Climbs the green stem, and slays the sleeping flower.

* Darwin believed in the spontaneous generation of life when conditions were suitable and nutrients available. This idea was out of favour for more than a century, but is now gradually coming back into discussion as a result of experiments on the origin of life.

Fell Oestrus* buries in her rapid course
Her countless brood in stag, or bull, or horse;
Whose hungry larva eats its living way,
Hatch'd by the warmth, and issues into day.
The wing'd Ichneumon for her embryon young
Gores with sharp horn the caterpillar throng.
The cruel larva mines its silky course,
And tears the vitals of its fostering nurse.
While fierce Libellula† with jaws of steel
Ingulfs an insect-province at a meal. . . .

Yes! smiling Flora drives her armèd car
Through the thick ranks of vegetable war;
Herb, shrub, and tree, with strong emotions rise
For light and air, and battle in the skies;
Whose roots diverging with opposing toil
Contend below for moisture and for soil;
Round the tall Elm the flattering Ivies bend,
And strangle, as they clasp, their struggling friend. . . .
And insect hordes with restless tooth devour
The unfolded bud, and pierce the ravell'd flower.

In ocean's pearly haunts, the waves beneath
Sits the grim monarch of insatiate Death;
The shark rapacious with descending blow
Darts on the scaly brood, that swims below;
The crawling crocodiles, beneath that move,
Arrest with rising jaw the tribes above;
With monstrous gape sepulchral whales devour
Shoals at a gulp, a million in an hour.
—Air, earth, and ocean, to astonish'd day
One scene of blood, one mighty tomb display!
From Hunger's arm the shafts of Death are hurl'd,
And one great Slaughter-house the warring world!
(IV 17–38, 41–8, 53–66)

The realization that Nature was 'red in tooth and claw' came as a nasty shock to many Victorians: but Darwin had

* Gadfly, bot-fly or sheep-fly.
†Dragonfly.

known about it long before, and had already proceeded one step further. He saw the inevitable effects of prolific over-breeding, but he also saw how it could add to what he calls the 'sum total of terrestrial happiness':

Each pregnant Oak ten thousand acorns forms
Profusely scatter'd by autumnal storms;
Ten thousand seeds each pregnant poppy sheds
Profusely scatter'd from its waving heads;
The countless Aphides, prolific tribe,
With greedy trunks the honey'd sap imbibe;
Swarm on each leaf with eggs or embryons big,
And pendent nations tenant every twig. . . .
—All these, increasing by successive birth,
Would each o'erpeople ocean, air, and earth.

So human progenies, if unrestrain'd,
By climate friended, and by food sustain'd,
O'er seas and soils, prolific hordes! would spread
Erelong, and deluge their terraqueous bed;
But war, and pestilence, disease, and dearth,
Sweep the superfluous myriads from the earth. . . .
The births and deaths contend with equal strife,
And every pore of Nature teems with Life;
Which buds or breathes from Indus to the Poles,
And Earth's vast surface kindles, as it rolls!

Hence when a Monarch or a mushroom dies,
Awhile extinct the organic matter lies;
But, as a few short hours or years revolve,
Alchemic powers the changing mass dissolve;
Born to new life unnumber'd insects pant,
New buds surround the microscopic plant;
Whose embryon senses, and unwearied frames,
Feel finer goads, and blush with purer flames. . . .

Thus the tall mountains, that emboss the lands,
Huge isles of rock, and continents of sands,
Whose dim extent eludes the inquiring sight,
ARE MIGHTY MONUMENTS OF PAST DELIGHT;
Shout round the globe, how Reproduction strives

With vanquish'd Death—and Happiness survives;
How Life increasing peoples every clime,
And young renascent Nature conquers Time.
(IV 347–54, 367–74, 379–90, 447–54)

This idea of the happiness of organic life is expressed more fully in *Phytologia* and is an important part of Darwin's evolutionary philosophy.

All organized nature may be divided into stationary organizations, and locomotive organizations; the former of which are called vegetables, and the latter animals. All those parts of vegetables which are most nutritious to animals consist . . . of aliment secreted from the vegetable blood, and laid up in reservoirs for the future sustenance of their embryon or infant progeny; which reservoirs are plundered by locomotive animals, and devoured along with the progeny they were designed to support! Add to this, that the stronger locomotive animals devour the weaker ones without mercy. Such is the condition of organic nature! whose first law might be expressed in the words, 'Eat or be eaten!' and which would seem to be one great slaughter-house, one universal scene of rapacity and injustice!

Where shall we find a benevolent idea to console us amid so much apparent misery? . . .

It may first be observed, that the seeds of plants and the eggs of animals, when they have left the pericarp or uterus, and have not yet commenced their new growth upon the soil, or beneath the wings of the mother, exist in a torpid state, not possessed of sensitive life; and cannot therefore at this time be supposed to suffer pain, when they are destroyed by other animals; though those animals obtain pleasure from the activity into which their vascular systems are excited by the stimulus of the aliment thus supplied. . . .

Fourthly, as all animal existence must perish in process of time . . . it is so ordered that as soon as any organized being becomes less irritable and less sensible, and in consequence feeble or sickly, that it is destroyed and eaten by other more irritable and more sensible, and in consequence more vigorous organized beings; as insects attack the weaker vegetable productions in preference to the healthy ones; and beasts of prey more easily catch and conquer the aged and infirm, and the young ones are defended by their parents. By this contrivance more pleasurable sensation exists in the world, as the organized matter is taken from a state of less irritability and less sensibility, and converted into a state of greater; that is in other words, that the old organizations, whether stationary or locomotive ones, are transmigrated into young ones: whence it happened that before mankind introduced rational society, and conquered the savage world, old age was unknown on earth!

Finally, the aged and infirm, from their present state of inirritability and insensibility, lose their lives with less pain, and which ceases instantly with the stroke of death; insomuch that death cannot so properly be called positive evil, as the termination of good. . . .

Many of the materials which have been thus produced by the digestion and secretion of organized beings, and have given pleasure in their production, have been slow in their decomposition after the death of the creature; as the shells of fish were originally thus formed, and were left at the bottom of the ocean, till they became wonderfully accumulated, were afterwards elevated by submarine fires, and constitute at this day the immense rocks and unmeasured strata of limestone, chalk, and marble.

The strata which are incumbent on the calcareous ones, which consist of coals, sand, iron, clay, and marl, are all of them believed to have been originally the products chiefly of vegetable organization, whatever changes they have since undergone in the long progress of their decomposition; and that all those solid parts of the earth have been thus fabricated from their simpler elements by vegetable and by animal life, and have given pleasure to those organized beings which formed them, at the time of their production.

We hence acquire this sublime and interesting idea; that all the calcareous mountains in the world, and all the strata of clay, coal, marl, sand, and iron, which are incumbent on them, are MONUMENTS OF THE PAST FELICITY OF ORGANIZED NATURE!—AND CONSEQUENTLY OF THE BENEVOLENCE OF THE DEITY!

(*Phytologia*, pp. 556, 557, 559–60)

The sudden apparition of the Deity in the very last line may come as rather a shock: but Darwin was merely trying to

FIG. 5 (*above*). Erasmus Darwin's house at Lichfield, as it appears today. The house, where he lived from 1758 until 1781, is at the western end of the Cathedral Close, facing Beacon Street, and the spires of the cathedral can be seen on the left of the photograph. (Photo: *The Birmingham Post*.) FIG. 6 (*below*). Ferry across the Derwent constructed by Darwin to link his house in Full Street, Derby, with the orchard, 1789. From a drawing by his son Francis, who is shown as the child in the boat, with his mother.

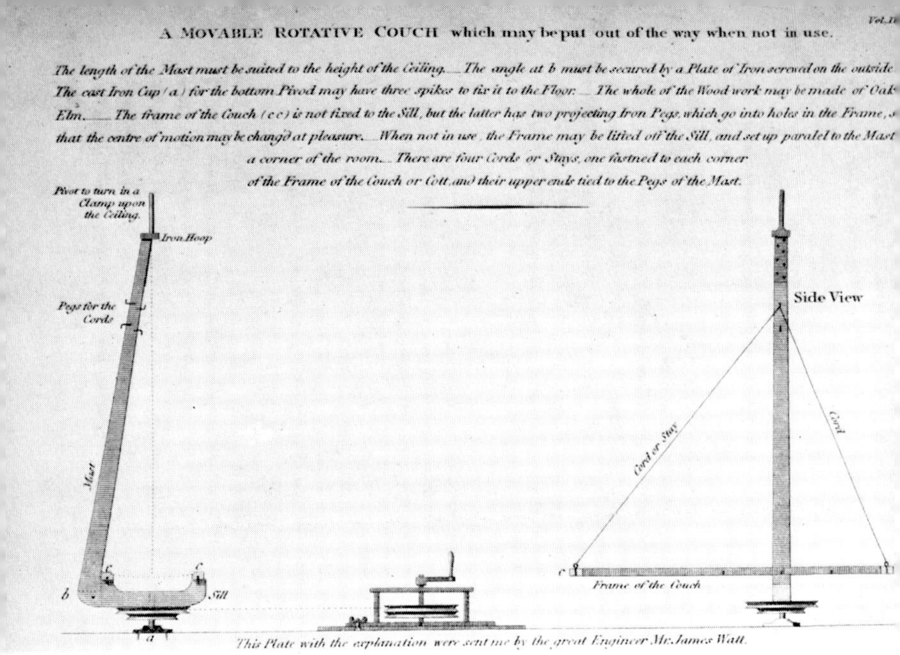

FIG. 7 (*above*). The 'movable rotative couch' designed by James Watt (about 1797) to try out Darwin's ideas on the effect of centrifugation. Though probably never constructed, the couch is the ancestor of the modern aero-medical centrifuge.

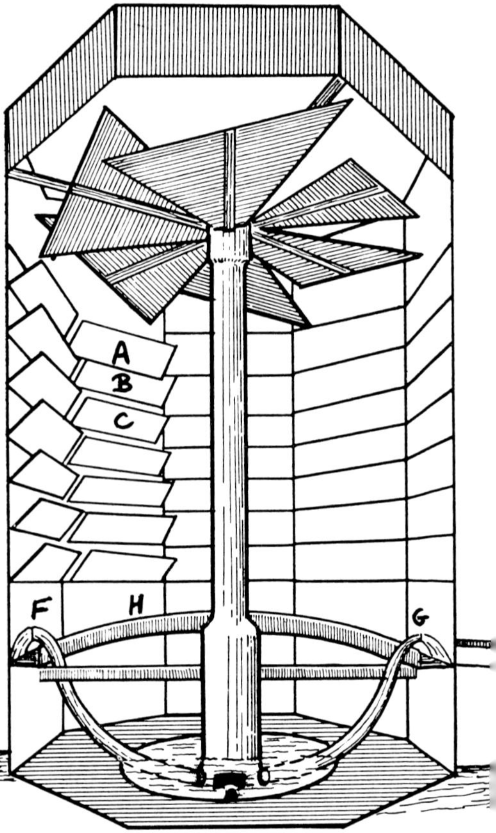

FIG. 8 (*right*). Darwin's horizontal windmill, re-drawn from Plate VII of *Phytologia*. The design shown is for raising water a few feet, as in the draining of morasses. An earlier version of the windmill, of which a model was made in 1769, was used in Wedgwood's factory at Etruria from 1778 until 1792 for grinding colours. See page 121 for an explanation of the mechanism.

avoid being attacked yet again for propounding atheistic ideas. He wished to be regarded as a Deist—a believer in a vague 'First Great Cause'. However, some of his alleged remarks about Christianity seem to take him well over half way along the road to atheism.

Darwin's evolutionary views were rejected in his own time, and Coleridge coined the scornful word *Darwinizing*, to mean 'making evolutionary speculations'. But in fact Darwin's ideas provide a remarkable anticipation of modern views. Particularly good are his pictures of the development from primitive molecules spontaneously generated, right up to the animals now existing, in a time-scale of 'millions of ages', i.e. hundreds of millions of years; and also his vivid realization of the struggle for existence. But perhaps even more impressive is the way he weaves his evolutionary ideas into the wider philosophy of organic happiness. For Erasmus Darwin evolution was no casual speculation, but a belief he lived with for thirty years and one which moulded his whole philosophy of life.

7 Plants in the Good Earth

> The fields his study, Nature was his book.
>
> R. BLOOMFIELD; *The Farmer's Boy*

DARWIN'S views on plant life are given in *Phytologia*, an impressive and surely-grounded survey of vegetable life and all that goes to sustain it, from carbon dioxide down to manures. Selections cannot really do justice to *Phytologia*: so many topics demand to be mentioned that little idea can be given of the attention to detail which is one of the book's great virtues.

Phytologia is in three parts:

(1) 'physiology of vegetation', with detailed accounts of plant structure and functioning (pages 1–139);

(2) 'economy of vegetation', covering seed growth, photosynthesis, nutrition, manures, drainage, aeration and diseases (pages 141–372);

(3) 'agriculture and horticulture', with the accent on productivity, of fruits, seeds, root-crops and flowers (pages 373–578).

The book is dedicated to Sir John Sinclair, President of the Board of Agriculture in the 1790s, and Darwin tells us why in his Introduction:

> Agriculture and Gardening, though of such great utility in producing the nutriment of mankind, continue to be only Arts, consisting of numerous detached facts and vague opinions, without a true theory to connect them, or to appreciate their analogy; at a time when many parts of knowledge of much inferior consequence have been nicely arranged, and digested into Sciences. . . .
>
> My inducement to commence this work, after it was suggested to me by the letters of Sir John Sinclair, was a belief that the experiments and observations already made on the growth of plants, with the modern improvements in chemistry, were

sufficiently numerous and accurate for the establishment of a true theory of vegetation. (p. vii, viii)

Darwin begins by discussing the individuality of buds:

> We have so accustomed ourselves to consider life and irritability to be associated with palpable warmth and visible motion that we find a renitency [reluctance] in ourselves to ascribe them to the comparatively cold and motionless fibres of plants. But to reason rightly on many vegetable phenomena we shall find it necessary first to show that vegetables are in reality an inferior order of animals.
>
> If a bud be torn from the branch of a tree, or cut out and planted in the earth with a glass cup inverted over it . . . or if it be inserted into the bark of another tree, it will grow, and become a plant in every respect like its parent. This evinces that every bud of a tree is an individual vegetable being; and that a tree therefore is a family or swarm of individual plants, like the polypus, with its young growing out of its sides, or like the branching cells of the coral-insect. (pp. 1–2)

Charles Darwin gave his grandfather credit for being the first to bring out clearly the principle of the individuality of buds.

Darwin goes on to discuss the 'absorbent vessels' of vegetables, which conduct the sap. Osmotic pressure, 'the force that through the green fuse drives the sap', had been measured by Hales in the 1720s, but its mechanism was not understood until the mid-nineteenth century. Darwin credits the absorbent vessels with the power to suck up sap—a somewhat feeble explanation. However, he rightly refutes the views of Malpighi and Grew that the absorbent vessels are merely air tubes, as in dead plants.

Next come seeds and growth. Darwin sees how sexual generation, as opposed to budding, provides the variety essential for improvement:

> This linear or paternal progeny of vegetables in buds or bulbs, or wires, is attended with a very curious circumstance, which is that they exactly resemble their parents when they are arrived at their maturity . . . as is observed in grafting fruit-trees, and in propagating flower-roots, or strawberries, or potatoes, by their wires or roots; whereas the seminal offspring of plants, as it derives its form in part from the mother as well as

> father, is liable to perpetual variation, both which events are employed to great advantage by skilful gardeners. (p. 26)

Next Darwin explains that leaves act as the 'lungs' of vegetables and are, like animal lungs, of enormous area. The veins in leaves are analogous to the network of blood pathways in the lungs, and he describes experiments he made in 1791 on a stalk of large spurge immersed in a decoction of madder, to show the 'veins and arteries' in red. He thought that the respiration (or transpiration as it would now be called) occurred through minute pores (now called stomata), a hypothesis he confirmed by experiment:

> As those insects which have many spiracula, or breathing apertures, as wasps and flies, are immediately suffocated by pouring oil upon them, in the year 1783 I carefully covered with oil the surfaces of several leaves of phlomis, of Portugal laurel, and balsams; and though it would not regularly adhere, I found them all die in a day or two, which shows another similitude between the lungs of animals and the leaves of vegetables (p. 45)

In section 5 of *Phytologia* he describes the 'aortal arteries and veins' and presses the analogy with animals too far. He believed that plants had a circulatory system rather than a 'one-way traffic' motion of the sap, thus returning to an error which Hales had exposed earlier in the century. This is the worst defect in Darwin's system of vegetable physiology, and it was partly responsible for his belief that vegetables enjoyed sensation, though in a much lesser degree than animals.

Section 6, on 'glands and secretions', is important because Darwin recognizes the conversion of starch and mucilage into sugar, and clearly brings out the vital importance of sugar, which was not appreciated in his day. He also, characteristically, remembers the slaves who harvested it:

> In like manner the digestive powers of the young vegetable, with the chemical agents of heat and moisture, convert the starch or mucilage of the root or seed into sugar for its own nourishment. . . . And thus it appears probable that sugar is the principal nutriment of both animal and vegetable beings. That it is the most nutritive part of vegetable substances is evinced by the well-ascertained fact that the slaves in Jamaica

grow fat in the sugar harvest, though they endure at that time much more labour. . . .

In many plants sugar is found ready prepared . . . thus in the beet-root, the crystals of it may be discerned by a microscope. . . . And in the sugar-cane it abounds in such large quantity as to contribute much to the nourishment of mankind. And—and what?—Great God of Justice! grant that it may soon be cultivated only by the hands of freedom, and may thence give happiness to the labourer, as well as to the merchant and consumer. (pp. 76–7)

Darwin also foresees that sugar will be economically produced in England from beet:

As all sugar is the same, when nicely cleaned, whether it be obtained from fruits, grains, roots, canes, or sap-juice, there is reason to believe that sugar as well as spirit may some time or other be economically procured from the vegetables of this climate, as Marggraf extracted it from the beet-root. (pp. 587–8)

Darwin was so keen to emphasize the nutritive power of sugar that he made a serious mistake about honey, or rather nectar, which he thought was re-absorbed by plants and was extremely nutritious to them. The visits of bees were therefore, he thought, very unwelcome to the plant, and he failed to notice the help they gave in fertilization—a strange lapse for the author of *The Loves of the Plants*, who was often accused of overemphasizing the importance of fertilization mechanisms.

His lapse is all the more ironic because the very next section deals with the reproductive organs of vegetables. Again Darwin notes the advantage of sexual reproduction as a source of improvement in species.

This degeneracy of trees or perennial herbaceous plants propagated by buds or root-scions is not I think to be ascribed simply to the age of the original seedling-tree, because each successive generation of buds or bulbs are as distinct from the parent, as the generation by seeds. But as the lateral progeny of vegetables have no source of improvement after they have arrived at their maturity, but are liable like other plants and animals to injuries from food and climate, which injuries produce hereditary diseases, it is to this circumstance that their degeneracy ought rather to be ascribed; whereas the sexual progeny of vegetables are liable to improvement by the intermixture of the individuals of the same, or even of different species to counteract the effects of hereditary diseases. . . .

From the sexual, or amatorial, generation of plants new varieties, or improvements, are frequently obtained; as many of the young plants from seeds are dissimilar to the parent, and some of them superior to the parent in the qualities we wish to possess. . . . Sexual reproduction is the chef d'œuvre, the master-piece of nature. (pp. 96, 115, 103)

He describes sexual reproduction in plants with plenty of accurate and convincing detail, e.g. on the methods of seed dispersal:

The various methods which nature has employed for the dispersion of seeds are worth the attention of the farmer and gardener, both for the purpose of preventing the growth of noxious seeds, and of collecting the profitable ones. The pericarp of some plants bursts with sudden violence when the seed is mature, and disperses it to considerable distance; as that of wood-sorrel, oxalis acetocalla; and of impatiens, touch me not. The seeds of many plants of the class syngenesia are furnished with a plume, by which admirable mechanism they are disseminated by the winds far from their parent stem, and look like a shuttlecock, as they fly. Other seeds are disseminated by animals; of these some attach themselves to their hair or feathers by a gluten, as mistletoe; others by hooks, as clivers, galium aperine; burdock, arctium lappa; hound's tongue, cynoglossum. Others are swallowed whole for the sake of the fruit, and voided uninjured, as the hawthorn, crataegus, juniper, and some grasses. And the seeds of aquatic plants, and of those which grow on the banks of rivers, are carried many miles by the currents into which they fall. (p. 113)

After noting the breeding of new hybrid varieties of animals, 'mules' he calls them, as well as plants, he advocates artificial insemination for improving breeds.

It is possible that many new kinds of mules, which might be useful for labour, or by their milk or wool, or for food, might still be produced by the method of Spallanzani, who diluted the seminal fluid of a dog with much warm water, and by injecting it fecundated a bitch, and produced puppies like the dog. (p. 119)

The next subject is the 'muscles, nerves and brain' of vegetables.

The various motions of peculiar parts of vegetables evince the existence of muscles and nerves in those parts, such as the closing of their petals, and calyxes, at the approach of night, or in cold or wet weather; though the fibres and nerves which constitute these muscles are too fine for anatomical demonstration.

Some vegetables fold the older leaves over the new buds at the extremity of their stalks during the night, as alsine, chickweed; others, as the mimosa, sensitive plant, fold the upper or polished sides of their leaves together during their sleep. The hedysarum gyrans whirls its leaves in various directions, when the air is still, by an apparently voluntary effort, probably for the purpose of respiration. The dionoea muscipula, Venus's fly-trap, closes its leaves from the stimulus of insects which crawl upon them, and pierces them with its prickles. And the apocynum androsemifolium contracts its petals or nectaries round the proboscis of the flies which stimulate it, and holds them till they die, or till the sleep of the plant releases them by the relaxation of its muscular action.

From these circumstances it appears that there are not only muscles about the moving foot-stalks or claws of the leaves and petals above mentioned; but that these muscles must be endued with nerves of sense as well as of motion. Now, as when one part of a leaf of mimosa is touched, the whole leaf falls, it follows that there must be a common sensorium, or brain, where the nerves communicate, belonging to this one leaf-bud. To evince this further another leaflet was slit with sharp scissors, and some seconds of time elapsed, before the plant seemed sensible of the injury; and then the whole plant collapsed as far as the principal stem. Afterwards a small drop of oil of vitriol was put on the bud in the bosom of a leaf of another sensitive plant; and, after about half a minute, when the brain of this bud could be supposed to be destroyed, the whole leaf fell, and rose no more. If the individual buds of plants possess muscles and nerves with a brain, or common sensorium, the following questions consequently occur, and should be answered in the

> affirmative. Have vegetable buds irritability? have they sensation? have they volition? have they associations of motion? I am persuaded they possess them all, though in a much inferior degree even than the cold-blooded animals. (pp. 132–3)

Darwin's conclusion that buds are, as it were, very inferior animals, provides the basis of his poem *The Loves of the Plants*, in which the activities of the pistils and stamens are personified. He also believes sleep is a voluntary action in plants. He thinks plants have sense organs: they perceive light and heat, and some, like mimosa, respond to touch:

> To these must be added the indubitable evidence of their passion of love, and of their necessity to sleep; and I think we may truly conclude that they are furnished with a brain or common sensorium belonging to each bud. (p. 139)

The weakest point in his thesis comes when he admits that he cannot decide where this alleged brain resides. However, as the quotations show, his idea of vegetable sensibility is well-argued, even if it is not very widely accepted today (despite those waggish road-signs, 'Danger: heavy plant crossing'). On this speculative note the first part of *Phytologia* ends.

Part II, 'Economy of Vegetation', begins with a forty-page account of how to grow seeds and bulbs and how to stimulate budding by judicious pruning. This 'gardener's handbook', excellent though it is, does not call for any comment or quotation here.

The next section, entitled 'Manures, or the food of plants', is a great advance on any previous analysis of plant nutrition. It extends to seventy-two pages and covers everything needed by a self-respecting plant, from carbon dioxide and water to phosphorus and nitrates. Darwin begins by consolidating his previous statement that sugar is the product of digestion in both animals and plants:

> In this process of digestion much sugar is produced, which is probably immediately selected and drank up by the numerous mouths of the lacteals or lymphatics. . . .
>
> In their infantine state therefore there is a wonderful analogy between plants and animals; and it is particularly curious to observe in the process of converting barley into malt by the

> germination of the seed, that the meal of the barley is in part converted into sugar by the digestion of the young plant exactly as in the animal stomach. The wonderful effect of vegetable digestion in producing sugar may be deduced from the great product of the sugar-cane, and of the maple-tree in America . . . and the wonderful effect of animal digestion in producing sugar appears in patients who labour under diabetes. (pp. 188, 189)

He also notes the differences between plants and animals, principally the animal's possession of a stomach where it can quickly digest food, which plants have to take more slowly; and also the plant's ability to live 'on water and air alone, with the carbonic acid' [i.e. carbon dioxide] and small amounts of other elements. Darwin recognizes that carbon dioxide and water provide the main food of plants, though he thinks the carbon dioxide has to be dissolved in water.

> This carbonic gas in its fluid state, or dissolved in water, not in its aerial or gaseous state, is the principal food of plants; as appears, because their solid fibres consist principally of carbon, and their fluids of water.
>
> Next to carbonic acid the aqueous acid, if it may be so called, or water, seems to afford the principal food of vegetables . . . . . . when vegetable leaves are exposed to the sun's light, they seem to give up oxygen gas; but in the dark they give up carbonic acid gas, like the breath of animals. (pp. 193, 194)

The process of photosynthesis which is the basis of all vegetable life may be summed up as:

$$\text{carbon dioxide} + \text{water} + \text{light energy} \rightarrow \text{sugar} + \text{oxygen}$$
$$(6CO_2 + 6H_2O + 673 \text{ calories} \rightarrow C_6H_{12}O_6 + 6O_2)$$

in the presence of chlorophyll. The quotations already given show how Darwin fully appreciated that carbon dioxide, water and light were the main items required, and that sugar and oxygen were the main products. Although he failed to summarize the process in the neat form of a chemical reaction, he nevertheless deserves the credit for first recognizing the balance sheet of photosynthesis, for he went a good deal further than Ingenhousz (1779), who had established that plants absorb carbon dioxide and give out oxygen in sunlight; and Darwin largely anticipated (by forty years) the work of Liebig, who usually receives credit for the discovery. The nearest Darwin comes to stating the reaction is when he says:

> by the decomposition of water in the vegetable system, when the hydrogen unites with carbon and produces oil, the oxygen becomes superfluous, and is in part exhaled. (p. 194)

Oil (i.e. hydrocarbons) is close to sugar (carbohydrate), but not quite close enough. Darwin also erred in overemphasizing the importance of the oxygen absorbed at night.

Photosynthesis provides the main source of nutriment for plants, but Darwin's recognition of other essential elements is even more remarkable, for many of the processes he expounds were not fully established until the second half of the nineteenth century. He begins with nitrogen:

> We now come to the other ingredient, which constitutes a much greater part of the atmosphere than the oxygen, and this is the azote, or nitrogen; which also seems much to contribute to the food or sustenance of vegetables; for though azote, or nitrogen, enters into animal bodies in much greater quantities perhaps than into vegetables, so as to constitute according to some chemical philosophers the principal difference between these two great classes of organized nature; yet it enters also into the vegetable system, and is given out by their putrefaction; and also when lime is applied to moist vegetables it disengages from them both hydrogen and azote, forming volatile alkali [ammonia]. . . . (p. 195)

Darwin received credit in his own day for recognizing the importance of nitrogen and the mechanism of its absorption, through trapping in the soil or the formation of ammonia.[27] But today most histories of science ascribe these discoveries to Liebig, who was born three years after *Phytologia* appeared.

Darwin also specifies the mode of formation of nitrates via nitrous acid (now called nitric acid):

> The azote of the atmosphere, when air is confined in the interstices of the soil newly turned over by the plough or spade, contributes to the production of the nitrous acid by its union with the oxygen of the atmosphere, with which it was before only diffused, or with the much greater source of oxygen from

> the decomposing water of the soil. At the same time another part of the abundant azote combines with the hydrogen of the decomposing water of the soil, and produces ammonia or volatile alkali; which contributes to the growth of vegetables many ways. (p. 196)

This is elaborated later, when Darwin discusses putrefaction:

> There also appears at the end of the putrefactive process to be a junction of azote with oxygen, producing the acid of nitre, which probably may contribute much to promote vegetation. This appears from the mode of procuring that acid in France and Prussia, and which might be successfully practised under every shed in our own farm-yards; as it consists in a due mixture of vegetable and animal recrements with soil, frequently turned over to expose it to the air, while it is defended by a shed from the sunshine and rain; which is thus at the same time adapted to produce the quickest vegetation, and to generate the nitrous acid. (pp. 232–3)

Darwin's recognition of the importance and one mode of production of nitrates again deserves comment, for it was a fact which Davy in his *Agricultural Chemistry* (1813) completely failed to bring out, and apparently nitrates were not found in the soil until 1856.[28] Though it had been known for centuries that nitre was good for soil, no useful analysis was possible until after nitrogen was isolated, in the 1770s.

Darwin continues with separate sections on each of the other nutrients. He emphasizes the enormous loss of water through 'perspiration' as he calls it, and the plant's consequent need for a continual supply.

Carbon is next on Darwin's menu for vegetables:

> When animal and vegetable bodies are burnt without the access of air, that is, when their volatile parts are sublimed; there remains a great quantity of charcoal, a much greater in vegetable bodies than in animal ones; this is termed carbon by the French school, when it is quite pure; and is now known to be one of the most universal materials of nature. And as vegetable bodies contain so much of it in their composition, they may be supposed to absorb it entire, where they grow vigorously. . . .

> The whole atmosphere contains always a quantity of it in the form of carbonic acid, or fixed air. (p. 201)

Darwin thought that plants absorb carbon from black earth as well as from carbon dioxide in the air and dissolved carbonates in limestone or chalk country.

From carbon Darwin passes to phosphorus:

> Another material which exists, I believe, universally in vegetables, and has not yet been sufficiently attended to, is phosphorus. This like the carbon, nitrogen, hydrogen, and sulphur, is probably a simple substance, as our present chemistry has not yet certainly analysed any of them; and therefore I suppose it is taken up entire by the absorbent vessels of vegetables, when it can be met with in a state of solution; though it may also be occasionally formed and secreted by them; and may hence be registered among the articles of their food or sustenance. (p. 207)

Sir John Russell, in his *History of Agricultural Science in Great Britain*, gives Darwin credit for first recognizing the importance of this element as well as nitrogen: 'This is the earliest statement by a British scientist I have seen that nitrogen and phosphate are essential for the growth of plants' (p. 63). Darwin's basic reason for believing phosphorus to be essential for plants was the tendency of decaying matter to phosphoresce, which he illustrates with increasingly vivid examples:

> When wood is decomposed by putrefaction in a certain degree of warmth and moisture, it is often seen to emit much light in dark evenings, when recently broken and exposed to the oxygen of the atmosphere, so as to alarm benighted passengers; which is undoubtedly owing to the phosphorus which it contains, and which is at this time converted into phosphoric acid. Such a light frequently is seen on putrescent veal, when kept in a certain degree of warmth and moisture; and on the sea-weed placed on the oysters packed in barrels, and sent into the country; and in the streets of Edinburgh, where the heads of the fish called whitings or haddies are frequently thrown out by the people, I have on a dark night easily seen the hour by holding one of them to my watch. (pp. 207–8)

Darwin then goes on to suggest that a search for calcium phosphate should be made in Britain—'we may hope by greater attention to discover a mountain of phosphate of lime in our own country' (p. 211). This was, at the lowest estimate, a remarkable precognition of the widespread search for phosphates later in the century. Calcium, too, he regards as a vital element, partly because lime combines with carbon to produce soluble carbonates.

After a useful but slightly irrelevant section on clay, we reach the mouldiest topic of all, manures provided by decomposing substances. First, there is the discussion of nitrates already noticed; then Darwin turns to bone dust. Sir John Sinclair gave him the credit for first explaining its use as a fertilizer, but Darwin himself makes no such claim and merely remarks that 'the grinding of bones to powder has already been applied to agriculture, and the chopping of woollen rags'. (p. 237).

The next section, on 'preservation of manures', is full of rich material and includes several original ideas. The first sentence states the theme:

> The fertility of all countries depends on the saving and using those kinds of matter which are fit for the reproduction of organized bodies. There is a proverb in China that for this purpose a wise man saves even the parings of his nails and the clippings of his hair. (p. 241)

Darwin strongly advises horizontal ploughing, to prevent soil being washed away in heavy showers, and then he proposes sewage farms:

> Secondly, the manures of towns and cities, which are all now left buried in deep wells, or carried away by soughs into the rivers, should be removed by a police, which is said to exist in China, and carried out of towns at stated intervals of time for the purposes of agriculture. (p. 242)

This proposal we have come to adopt. But Darwin's solution to the problem of grave overcrowding is even more radical.

> Proper burial grounds should be consecrated out of towns, and divided into two compartments, the earth from one of which, saturated with animal decomposition, should be taken

> away once in ten or twenty years, for the purposes of agriculture; and sand or clay, or less fertile soil, brought into its place. (p. 243)

Like his grandson Charles he was greatly impressed by the rise in ground level, which gives greater weight to his idea:

> A great rise of the soil, from the remains of the bodies entombed in it, is seen round the churches of almost all populous towns; so as to have rendered it necessary to descend by several steps into those churches which were originally built so as to require steps to ascend into them; as may frequently be seen by the base of the architecture. (p. 243)

Darwin's concern for manure reaches its finest moment in an admirable injunction he proposes:

> There is a provident adage of general benevolence, 'Burn nothing which any animal will eat'; that is, 'Burn nothing which may nourish animals by its digestion in their stomachs'. May not the same benevolent idea be extended to the vegetable world, and say, 'Burn nothing which may nourish vegetables by its slow decomposition beneath the soil, which constitutes their stomachs'. (p. 244)

The unnecessary burning of 'a hair or a straw' should therefore 'give some compunctions to a mind of universal sympathy'. (p. 255)

In this remarkable 25,000-word chapter on Manures, Darwin goes a long way towards a complete exposition of the two fundamental processes in nature—the carbon and nitrogen cycles. New historians of science would do well to read Darwin before committing themselves to the traditional belief that these processes were discovered much later. They might also compare Darwin's analysis of manures (1800) with Davy's, in his *Agricultural Chemistry* (1813): although Davy was incomparably the better chemist, his descriptive account of manures, with no indication of the importance of nitrogen or phosphorus, seems outdated and almost naive after Darwin's analytical treatment.

Draining and watering are the next topics to occupy Darwin. First he explains the functioning of springs: most hills were formed, he believes, by being raised from beneath.

Hence the strata which compose the sides of mountains lie slanting downwards; and one or two or more of the external strata not reaching to the summit when the mountain was raised up, the second or third stratum, or a more inferior one, is there exposed to-day. This may be well represented by forcibly thrusting a very blunt instrument through some folds of paper; a bur will be raised with the lowermost leaf standing highest in the center of it. Or if at the original elevation of an extensive mountain the lowest stratum should not at first stand higher in the center of the summit, it would in time become so by some of the upper strata of the mountain being gradually washed away by rains into the valleys or rivers. (p. 258)

The moisture condensing near the summits of hills

descends between the strata of the incumbent soil, sometimes for many miles together; but generally from the nearest eminences into the adjoining valleys. (p. 259)

Darwin offers examples from Derby and then from one of his former haunts.

But there is a situation where the manner of the production of springs is most agreeably visible; it is about a mile from the city of Lichfield, near the cold bath erected by Sir John Floyer, in a beautiful piece of ground, which was formerly Dr. Darwin's botanic garden.

In this place a grotto about six yards wide and ten long has been excavated on the side of a hill consisting of siliceous sandstone with this peculiar circumstance; that the upper stratum of the sand-rock, which is there about five feet thick, is divided from the lower stratum of it by a sheet of clay not more than three or four inches in thickness; on the upper surface of this sheet of clay, between the lips of these rocks, a perpetual dribbling of water oozes quite round the grotto, like a shower from a weeping rock. Such sheets of water having been often observed to slide between the strata of the earth almost horizontally, like the horizontal joints of a stone-wall, have, I suppose, given the name of wall-springs to them, to distinguish them from pipe-springs, or such as burst out in a single rill. (p. 260)

To drain an alluvial valley, Darwin recommends digging a ditch (or boring holes) at the edge of the valley and perpendicular to the strata, rather than vertically, to intercept water running down the hillside.

It was Darwin who gave the first full and correct explanation of artesian wells (in the paper describing the well in his own garden at Derby, see Chapter 12); and perhaps the most influential section of *Phytologia* was that advocating the wider use of artesian wells for providing water supplies.

> The knowledge of this part of geology concerning the formation of springs may be employed for many useful purposes. . . .
>
> Thus where water, descending in high columns between the strata of mountains, is dammed up below by the materials which fill up the valleys; if a hole be bored in the valley deep through the incumbent soil and strata, [the water] frequently rises much above the source of the new aperture, and sometimes above the surface of the ground. . . .
>
> In the town of Richmond in Surrey, and at Inskip near Preston, in Lancashire, I am informed that it is usual to bore for water to a certain depth; and that when it is found in both these places, it rises so high as to flow over the surface. And there is reason to conclude that if similar experiments were made in many other places, such artificial springs might be produced at small expense, both for the common purposes of life, and for the great improvement of lands by watering them. (pp. 265–6)

Darwin's suggestions have been fully justified by subsequent events.

'The knowledge of this part of geology leads to another useful purpose, the discovery of springs'. In searching for springs,

> attention should be given to the inclination of the strata of that part of the country, which may be often seen in marl-pits, gravel-pits, or in hollow lanes. . . .
>
> A second observation for the purpose of detecting springs may be made on misty evenings; as those parts of the ground where the mist commences are moister than those in their vicinity on the same level; and in consequence may generally,

> if they are not hollow basins, possess springs nearer the surface; for these moister parts of the ground, having evaporated more during the day, are become colder on their surfaces than the drier ground in their vicinity; and in misty evenings, which are the same time calm, the stationary air over these moist parts of the ground is also more loaded with the evaporated moisture; and on both these accounts these moister situations are liable to show a condensation of aerial vapour sooner than other places on the same level. . . .
>
> Another observation for detecting springs may be made in rimy mornings; for as moist earth is a better conductor of heat than dry earth, the rime will sooner melt on those parts of the soil which are kept moist by springs under it than on other parts; as the common heat of the earth, which is 48° in this country, will sooner be conducted upwards in most places to dissolve the rime on the surface. On this account the rime is frequently seen on frosty mornings, when the heat of the air is not much above 32°, to lie an hour longer on dry cakes of cow-dung, or on bridges, or planks of wood, than on the common moist ground; as the latter much better conducts the common heat of the earth to the incumbent rime, which is in contact with it. (pp. 267–8)

After telling us how to find water, Darwin explains how to get rid of it, particularly from morasses. Since he cannot play tricks with the strata he settles for traditional methods, with ditches, dykes, pumps and siphons. He thinks it would benefit the national economy to drain all fens—

> But this, I fear, is an effort not to be expected in the present times, when the enclosure of forests and large commons is prevented by the interest of individuals, or by the difficulty of procuring expensive acts of parliament for every minute district, instead of including them in a general act, so meritoriously contended for by Sir John Sinclair. (pp. 271–2)

Darwin explains the advantages of periodically flooding meadow lands, and ends the chapter with two inventions of his own for raising water, a water-machine and a horizontal windmill. These are described in Chapter 8.

After so much water we come up for air, in the form of a chapter on the aeration and pulverization of soil.

As almost all soils not only contain carbon, and other inflammable materials, which are capable of uniting with oxygen, and thus producing the carbonic and other acids; but also contain water, which by its decomposition, when in contact with confined air, produces ammonia or volatile alkali by the union of its hydrogen with azote; and nitre by the union of its abundant oxygen with another part of the abundant azote or nitrogen of the atmospheric air; there is reason to conclude that the great use of turning over the soil with the plough or spade depends principally in the production of these effects by the confinement of both the oxygen and the azote or nitrogen of the air in the interstices of the soil; and on this account we have entitled this section the aeration of the soil rather than the oxygenation of it. (p. 283)

After this admirable introduction Darwin praises Tull's drill plough, with its advantages of even sowing and depth. However he criticizes Tull's machine for not dispensing seed uniformly, and, ever ready to help, offers a machine of his own invention, described fully in the Appendix to *Phytologia*. The detailed instructions and engineering drawings are too lengthy and complex to reproduce here; but this is obviously one of his most carefully thought-out inventions, as the first paragraph of the Appendix indicates:

The first experiment I tried to improve this valuable machine was . . . by enlarging the axis of Mr. Tull's seed-box into a wheel of sixteen inches diameter, with excavations in the rim to raise portions of the corn above the surface of that in the seed-box. But I found to my surprise the friction of the corn to be so much greater than expected, when six such large wheels were immersed in it, that an additional hopper became necessary to deliver the seed slowly into the seed-box, as in Mr. Cook's drill plough; which, as it would add much to the intricacy and expense of the machinery, and to the inaccuracy of the quantity of seed delivered, occasioned me to relinquish that idea, and after many designs and many experiments to construct the following machine, which I believe to be more simple, and consequently less expensive to construct, and less liable to be out of order, and to deliver the seeds of all kinds with greater accuracy than any drill plough at present in use; and that it possesses every other advantage that they can boast. The scale

> of the three following plates is half an inch to ten inches. . . . (p. 597)

The drawings which follow are accompanied by fourteen pages of detailed instructions.

Part II of *Phytologia* ends with a long discourse on diseases, ranging from 'diseased irritability', i.e. exhaustion, to mildew, blight, ergot, canker and other loathsome fungoid growths. Insects are another pest, and Darwin has a long account of the sap-sucking aphis (or greenfly), which breeds asexually in spring for nine generations and then in about October produces males and females for the final batch of eggs, laid ready to be hatched by the next year's vernal sun. For flies and wasps Darwin had earlier prescribed a scientific insecticide known in antiquity, namely

> a slight solution of arsenic, with a little sugar, put thinly on a plate or two, and set on chimney-pieces or windows. (p. 81)

The best conventional insecticide he can offer for the aphis is tobacco smoke or 'the fine powder of tobacco called Scotch snuff' applied with a powder puffer.

But he has another trick up his sleeve.

> The most ingenious manner of destroying the aphis would be effected by the propagation of its greatest enemy, the larva of the aphidivorous fly . . . which is said to deposit its eggs where the aphis abounds; . . . as soon as the larvae are produced, they devour hundreds around them with the necessity of no other movements but by turning to the right or left, arresting the aphis and sucking its juices. If these eggs could be collected and carefully preserved during the winter, and properly disposed on nectarine and peach-trees in the early spring, or protected from injury in hot-houses, it is probable that this plague of the aphis might be counteracted by the natural means of devouring one insect by another. (p. 356)

Darwin then gives a detailed description of this aphidivorous insect, the syrphid fly, with diagrams: he also advocates breeding the ichneumon fly, which eats caterpillars. The idea of scientifically-regulated biological control of insects was new, and Darwin is recognized as the pioneer in this important subject.[29]

The final 'disease' of plants is a very damaging one—being eaten by vermin. So we are told how to trap mice, rats and moles, by both traditional and novel methods. One of Darwin's brightest ideas is to apply biological control to rats by importing diseases:

> In the Transactions of the American Philosophical Society there is a paper showing that the water-rats of that part of the country are so liable to be affected with tape-worm as is supposed much to diminish their numbers. In this country many animals, as I believe dogs, cats, and geese, as well as the human species, are afflicted with this intestine enemy. Could some of these diseased American rats be imported into this country, and propagate their malady amongst the native rats of this climate? (pp. 583–4)

Darwin must therefore be reckoned a pioneer in germ-warfare, or at least worm-warfare, and his idea was recently paralleled, half intentionally, with myxomatosis and rabbits.

Part III of *Phytologia* is even more practical than Part II, being concerned with food production. The first section, on Fruits, runs to over fifty pages, full of ripe wisdom about pruning, grafting, preservation, etc. Darwin goes into great detail, where we cannot follow him. His advice for perfecting the fruit—thinning out fruits which are crowded, protecting from frost, etc.—is sound but unremarkable. The most unusual twist comes at the end when, 'to amuse the reader', Darwin turns some of his instructions into verse:

> Behead new-grafted trees in spring,
> Ere the first cuckoo tries to sing;
> But leave four swelling buds to grow
> With wide-diverging arms below;
> Or fix one central trunk erect,
> And on each side its boughs deflect.
>   In summer hours from fertile stems
> Rub off the supernumerous gems;
> But where unfruitful branches rise
> In proud luxuriance to the skies,
> Exsect the exuberant growths, or bind
> A wiry ringlet round the rind. . . . (pp. 429–30)

Fifteen further couplets are available for any reader not yet amused enough.

Darwin is just as keen on the production of seeds, chiefly because ground yields more food when planted with grain than when under animal pasturage. Again he goes into great detail over the choice of seeds and the methods of sowing them. As we might expect, he is against one use of grain, and he suggests

> a total prohibition of the destructive manufactory of grain into spirits, or into strong ale, and thus converting the natural nutriment of mankind into a chemical poison. . . . (p. 468)

Though he prefers grain crops to pasturage, Darwin was no vegetarian:

> Mankind nevertheless seems by nature to be designed to subsist on both vegetable and animal nutriment, which appears from the length of his intestines, which like those of swine are much longer than the intestines of carnivorous animals, and much shorter than those of the vegetable eaters; and which also appears from the structure of his teeth, which partakes of the structure of those of the carnivorous and phytivorous animals. (p. 469)

Next Darwin gets down to the roots, with some intriguing examples, including a new variety of potato growing in the odorous garden of a well-named military friend:

> I was this day shown by my friend Major Trowel of Derby a new variety of the potato in his excellent new-made garden, the soil of which consists of marl mixed with lime and stable-manure. From one root there appeared to issue six or eight stems three or four feet long, at every joint of which were produced new potatoes; at the lower joints there were three of these aerial potatoes, one large, one the size of a pullet's egg, and a smaller one on each side of it. . . . I should hope this proliferous variety by cultivation may become permanent, and give rise to a new species, which may produce both aerial potatoes and subterraneous ones, a twofold viviparous progeny. (pp. 474, 475)

Today we do not often hear about the barks of trees. But Darwin has a good deal to say on the subject, for in his day barks, and especially the Peruvian bark (quinine), were a valued item in the materia medica, and were used for many other purposes, e.g. as structural materials and for dyeing.

From barks Darwin proceeds to leaves and wood, and in particular to grasses. He recommends the best varieties for various situations and teaches 'the art of making hay'—to evaporate about two-thirds of its weight. Another art he discusses is etiolation, i.e. excluding light to make the plant pale, practised on celery and young ladies:

> A seclusion from the sun's light and from air has an effect somewhat similar on animal bodies, rendering them pale and weak, as may be seen in the etiolated young ladies of some boarding schools. (p. 511)

Darwin is just as confident and detailed when he discusses the art of the woodman: reading through *Phytologia* makes you realize what Coleridge meant when he said Darwin had 'a greater range of knowledge than any other man in Europe'. When faced with crooked young trees, for example, he advises scratching them on the concave side to make them grow straight. The best way of growing timber straight and true, he says, is to plant the trees close together, because

> the powerful contest with each other for light and air propels them upwards, instead of producing many lateral branches. (p. 516)

Phytologia was written in the last few years of the eighteenth century and the growing threat of a French invasion threw its shadow even on the trees.

> The political advantage or disadvantage of cultivating timber in this island should here be considered. In the present insane state of human society . . . war and its preparations employ the ingenuity and labour of almost all nations; and mankind destroy or enslave each other with as little mercy as they destroy and enslave the bestial world. . . .
>
> Now as the power to resist invasion, and to defend our natural rights, when infringed by foreign enemies, must depend more on the number of men than on the number of trees; there need be no hesitation in determining that those lands, which can be employed in the present production of vegetable or animal food,

> should not be occupied in the tedious cultivation of future timber.
>
> But . . . the summits of this country consist principally of a ridge of mountains extending from south to north between the eastern and western seas, as those of the Peak of Derbyshire and the Moorlands of Staffordshire, which are so bleak or so barren as to be totally unfit for the plough or for pasturage, and yet might be employed for raising variety of timbers. . . . All those unfertile mountains from the extremity of Cornwall to the extremity of Scotland, should be covered with extensive forests of such kinds of wood as experience has shown them to be capable to sustain. (pp. 526–7, 527–8)

What kinds of wood? 'Certainly pines, as Scotch fir, might in these situations ["barren mountains"] succeed astonishingly' (p. 517). Thus Darwin completely specifies the afforestation programme put into operation by the Forestry Commission in the twentieth century.

Finally we come to flowers, valued for their beauty rather than utility. Darwin's detailed instructions on the best ways of growing a beautiful flower-garden are most felicitously rounded off with the discourse on the happiness of organic life already quoted in Chapter 6. The fine crescendo of this section would have provided a fitting climax to *Phytologia*. But Darwin thought our organic happiness would be further increased by reading a long poem about broccoli, translated in part from 'an elegant Latin poem of Edward Tighe, Esq.' Perhaps he was right, for this mock-heroic instruction-manual on broccoli-growing is one of the funniest of the poems he inserted 'to amuse the reader'. He begins with a solemn invocation to the rural Muse:

> This boon I ask of Fate, where'er I dine,
> O, be the Proteus-form of cabbage mine!—
> Cale, colewort, cauliflower, or soft and clear
> If Broccoli delight thy nicer ear,
> Give, rural Muse! the culture and the name
> In verse immortal to the rolls of Fame. (p. 560)

He ends with good wishes to the author:

> Oft in each month, poetic Tighe! be thine
> To dish green Broccoli with savory chine;

Oft down thy tuneful throat be thine to cram
The snow-white cauliflower with fowl and ham! (p. 563)

This is a happy note on which to end our sampling of *Phytologia*, a book that deserves to be recognized as a classic in the development of ideas on plant nutrition and agricultural chemistry, if only for its nearly complete description of photosynthesis and its correct catalogue of the chemical elements that are essential to plants. But of course *Phytologia* has much more: engineering drawings for the drill-plough and pumps; the proposals for sewage farms, afforestation of highlands with firs, and a system of artesian wells; the improvement of animal breeds by artificial insemination; the biological control of insect pests by breeding predators; the reform of cemeteries and the philosophy of organic happiness—not to mention Major Trowel's aerial potato and the poem on broccoli.

8 Inventions Galore

> The most inventive of philosophical men.
>
> COLERIDGE on Darwin

DARWIN'S passion for invention emerges most forcefully in his unpublished Commonplace Book, which has dozens of designs for mechanical inventions. Some of these machines were built under his direction, and were proved practical; some remained as sketches, but were later re-invented and found to be workable; and some are mere curiosities.

The horizontal windmill was probably the most useful of Darwin's inventions. A windmill of his design was at work grinding colours in Wedgwood's factory at Etruria from 1778 until 1792, when it was displaced by steam. But Darwin also intended the windmill for use as a pump to drain morasses or to irrigate land a few feet above the water table. This is the version illustrated in *Phytologia* and re-drawn in Fig. 8. Darwin's description of its working, somewhat shortened, is as follows:

> It consists of a windmill sail placed horizontally like that of a smoak-jack, surrounded by an octagon tower. . . . [The pillars at the 8 corners] are connected together by oblique horizontal boards A, B, C . . . at an angle of about 45° . . .; the wind as it strikes against any of them, from whatever quarter it comes, is bent upwards and then strikes against the horizontal wind-sail. These horizontal boards, which form the sides of the octagon, may either be fixed in their situations or be made to turn upon an axis a little below their centres of gravity, so as to close themselves on that side of the octagon tower most distant from the wind.
>
> It may be supposed that the wind thus reflected would lose considerably of its power before it strikes on the wind-sail, but [experiments have shown that this is not so]. . . . There is reason to conclude that the power of this horizontal wind-sail may be considerably greater than if the same sail was placed nearly vertically opposed to the wind in the usual manner.

> At the bottom of the shaft of the wind-sail is placed a centrifugal pump with two arms at F, G. . . . It consists simply of an upright bored trunk, or cylinder of lead, with two opposite arms with an adapted valve at the bottom to prevent the return of the water, and a valve at the extremity of each arm to prevent any ingress of air above the current of the water as it flows out.
>
> H is a circular trough to receive the streams of water from F and G, to convey them where required. (*Phytologia*, Plate VII, after p. 282; the lettering on the diagram has been altered.)

The experiments Darwin mentions are recorded in his Commonplace Book in 1779. First he tried the rotor naked, and found it turned 875 times in 84 seconds. Then he enclosed it in a square box with two sides taken off, so as to present one corner to the wind, and the number of turns decreased to 820. Next he added a horizontal board, which caused a further decrease, to 804. Finally he added little wings and a board underneath, and managed to obtain 1,053 turns. In a footnote, he remarks 'N.B. I made all these experiments with great accuracy, such as was quite satisfactory to myself'.

Darwin's first designs for the windmill had been made in the 1760s, when Wedgwood wanted the mill for his factory at Etruria. In 1769 Darwin wrote to the Society of Arts:

> I have lately constructed the model of an horizontal Wind-mill, which appears to have a third more Power than any vertical Wind-mill, whose Sail is of the same diameter, and is in other respects more manageable and less liable to Repair, as it has less wheel-work, having only an upright Shaft on the Top of which the Sail is fix'd. . . . This Model is 3 ft. diameter, and I should not chuse the Expense of sending it up unless there was *a Chance* . . . that the Society might grant me a Premium *to assist* me to execute it at large.
>
> (Letter to Dr Templeman, 4 Feb. 1769. R. E. Schofield, *Lunar Society*, p. 74)

When poets refer to the wind of inspiration, blowing where it listeth, they are being metaphorical: but Darwin's inventiveness was literally inspired by wind, for his great interest in the atmosphere led him to design various wind-gauges, including an ingenious machine to measure the net flow of air from north to south, shown in Fig. 9. His explanation of this device is as follows:

> A machine to measure the quantity of air passing from N to S and from S to N is not impossible to construct. Let there be a tube about six inches diameter and a foot or ½ a yard long, placed exactly in the meridian on a chimney with a windmill sail in the center of it, having a crank which at every revolution should strike against the extremity of a rod which might come down into a garret and be made to register the number of revolutions of the fly [i.e. rotor] for a month or two together one way, and to unwind again when the wind blew in the contrary direction. (Commonplace Book, p. 46)

In the diagram drawn by Darwin, Fig. 9, the tube is more slender than he implies in his description. The line of circles at the bottom of the diagram represents reduction gearing, the rotation of the cog-wheel furthest to the left being perhaps 100,000 times slower than that of the primary wheel on the right—rather like a household electricity meter today.

Darwin was particularly interested in the weather because he met its full fury on his daily rounds and had to make his own forecasts of whether his carriage would get stuck in the mud. One of his aids in weather forecasting was a pointer on the ceiling of his study, which was connected to a weathercock on the roof above and gave him a continuous indication of the wind direction. This was an early but effective example of continuous telemetry, and by monitoring the progress of the winds from his armchair, Darwin discovered the importance of cold and warm fronts, as we shall see in Chapter 12. Meteorologists were remiss in not following up Darwin's example of how to observe in comfort: for too long they made a fetish of intermittent on-the-spot observations and did not try continuous telemetry.

Another wind-operated gadget, though of a very different kind, was his speaking-machine. This is perhaps the most impressive of all Darwin's inventions, because he first had to develop a theory of phonetics. He divides the sounds of speech into four classes: (1) clear continued sounds, vowels; (2) hissing sounds, sibilants; (3) semi-vocals, a mixture of (1) and (2); and (4) interrupted sounds, or consonants. He analyses the vibrations of the air which give rise to each. The vowels, for example,

> are produced by the streams of air passing from the lungs in respiration through the larynx; which is furnished with many small muscles, which by their action give a proper tension to the extremity of this tube; and the sounds, I suppose, are produced by the opening and closing of its aperture; something like the trumpet stop of an organ, as may be observed by blowing through the wind-pipe of a dead goose. . . .
>
> The hissing sounds are produced by air forcibly pushed through certain passages of the mouth without being previously rendered sonorous by the larynx; and obtain their sibilancy from their slower vibrations, occasioned by the mucous membrane, which lines those apertures or passages, being less tense than that of the larynx. . . .

and so on, with much detail. He then proceeds to describe the exact formation of each sound:

> P. If the lips be pressed close together and some air be condensed in the mouth behind them, on opening the lips the mute consonant P begins a syllable. . . .

And similarly for every other sound. He found most difficulty, he says, with the vowels, especially in deciding where exactly in the mouth they were modulated.

With this wealth of theory behind him, he embarked on the construction of his speaking-machine, which he describes as follows:

> I have treated with greater confidence on the formation of articulate sounds, as I many years ago gave considerable attention to this subject for the purpose of improving shorthand; at that time I contrived a wooden mouth with lips of soft leather, and with a valve over the back part of it for nostrils, both which could be quickly opened or closed by the pressure of the fingers; the vocality was given by a silk ribbon about an inch long and a quarter of an inch wide stretched between two bits of smooth wood a little hollowed; so that when a gentle current of air from bellows was blown on the edge of the ribbon, it gave an agreeable tone, as it vibrated between the wooden sides, much like a human voice. This head pronounced the *p*, *b*, *m*,

> and the vowel *a*, with so great nicety as to deceive all who heard it unseen, when it pronounced the words *mama*, *papa*, *map*, and *pam*; and had a most plaintive tone, when the lips were gradually closed. My other occupations prevented me from proceeding in the further construction of this machine; which might have required but thirteen movements, as shown in the above analysis, unless some variety of musical note was to be added to the vocality produced in the larynx; all of which movements might communicate with the keys of a harpsichord or forte piano, and perform the song as well as the accompaniment; or which if built in a gigantic form, might speak so loud as to command an army or instruct a crowd.
> (Quotations from *The Temple of Nature*, Additional Note No. XV, pp. 107–120)

His speaking-machine created a sensation when it was first heard, but unfortunately no picture of it has survived. In 1771 Boulton, in a document witnessed by Keir and Small (see Fig. 10), promised to pay Darwin £1,000 for a machine 'capable of pronouncing the Lord's Prayer, the Creed, and Ten Commandments'. However, this may have been a joke against the anti-Christian Darwin.

From this precursor of the phonograph we pass to a precursor of the guided missile, in the form of a design for an artificial bird with flapping wings, Fig. 12. The explanation to go with this is as follows:

> Let a watch-spring be fixed with one end to a frame and the other wrapped round an axis; at each end of the axis let a wheel be put with teeth of such form and situation that they shall move a wing, like a bat's wing, or like a ladies fan, one tooth carrying it downwards, another carrying it towards the body, another carrying it upwards and a fourth outwards again from the body. N.B. One edge of the wing is to be fastened to the body and the other to a kind of fan-stick made of a porcupine quill. The tail of feathers spread out and lying obliquely to the action of the wings, or rather to its intended track in the air. (Commonplace Book, p. 33)

Then, a few pages later, comes the excellent idea of replacing the watch-spring by a compressed-air bottle, the very method often used for actuating the control surfaces of modern guided missiles:

> But the simpler way, and better, would be to have a copper globe fill'd with condensed air, and the tube to make the longest feather in the tail of the bird. (Commonplace Book, p. 38)

Thus Darwin succeeded in overcoming the main bugbear of early aircraft designers—the lack of power. But I fear his feathery flapper would have been aerodynamically unstable.

His up-to-date propulsion for aeronautics is matched by what may be up-to-date propulsion for astronautics, namely a very rough sketch, without explanation, of what appears to be a rocket motor with hydrogen and oxygen as propellants, though diluted by ordinary air. He shows the hydrogen and oxygen stored in separate containers and being fed into a cylindrical combustion chamber with an exit nozzle at one end, rather like the propellant feed and combustion chamber of a modern rocket. Tsiolkovskii (1903) is usually credited with the idea of the hydrogen-oxygen rocket, which is being used in the upper stages of the United States Apollo project for placing men on the Moon.

From astronautics we can proceed to astronomy, for Darwin devotes much attention to telescopes made up of a number of small lenses or mirrors.

> Suppose twenty glasses, either lenses or concave specula, are so placed as to throw all their images of a certain object on to one focus—there will then be one image with 20 times the brightness. (Commonplace Book, p. 80)

So far this idea has not been much used; but now that lasers may allow very accurate positioning of the mirrors, it becomes possible to contemplate an optical telescope of Darwinian design with an aperture considerably greater than that of the 200-inch telescope at Mount Palomar. The mirrors of telescopes carried on space vehicles may also be constructed on the Darwinian principle.[30]

The next invention is also concerned with producing clear images, not of distant stars but of mundane documents: it is one of the earliest office copying-machines. In Darwin's day documents were copied by hand: he saw the advantages of a mechanical copier and in about 1777 he devised his 'bigraphers', one of which is shown in Fig. 11. Darwin's design apparently worked quite well, though rather large and clumsy. The words written with the quill on the right are reproduced simultaneously by the tube in the middle, although, as Darwin remarks,

> the letters made by the tube will be $\frac{1}{4}$ thinner or narrower than those of the pen, but the same height. (Commonplace Book, p. 52)

Darwin's success was short-lived, however, for his invention provoked Watt to seek a better method, which he soon found: 'I have fallen on a way of copying writing chemically, which beats your bigrapher hollow', Watt wrote triumphantly in 1779. Watt's chemical process remained the standard copying method, used in offices for nearly a hundred years. The modern descendant of Watt's machine is the xerographic copying-machine, which depends on electrostatic processes.

This brings us back to Darwin again, for he was a keen inventor of electrostatic generators. His 'electrical doubler' was quite famous in its time, and Fig. 13 shows a sketch of it in his Commonplace Book (p. 79), slightly touched up. The central disc, apparently of glass with a metal sheet imbedded in it, was given an initial charge. The two outer discs are 'movable brass plates which being mounted on cranked levers, could be made to approach or recede from the central plate by turning the winch'.[31] As the plates approach the central one, charges are induced on them which apparently pass to a Leyden jar via the conductor *w*. Then the plates are withdrawn and earthed via *e* and *z*. The mechanism of Darwin's rough sketch is not very clear, but apparently the design worked.

Many of Darwin's inventions sprang from the needs of his everyday life. The most pressing need was for a safe and comfortable carriage to take him on his rounds. It was for his carriage designs that he won most renown as an inventor, but unfortunately no sketches of them seem to have survived: all that we have are some unsatisfactory second-hand verbal descriptions. One of Darwin's favourite designs 'was in the form of a garden chair, fix'd upon the axle-tree; the footboard terminated in an elastic pole', which was connected to the top of a short vertical iron bar fitting into a socket in the saddle, so that the horse was free to move in any direction it wished. The horse was not always co-operative, however, and one day in 1768, after taking fright, it gave the carriage a violent shock. The very next day the axle broke and Darwin was pitched out, fracturing his knee-cap.[32]

The suspension and springing of carriages—what might be called the undercarriage—also exercised Darwin's ingenuity. As early as 1754 he had the idea of using the wheel spokes as springs, and a coach constructed on these principles, with each spoke curved into the serpentine line of beauty, was made and greatly admired, especially by Wedgwood.

An even smoother way of leaving home was provided by the ferry across the river Derwent at Derby between his house and the orchard. The sketch of the ferry by his son Francis, Fig. 6, does not make the mechanism entirely clear; but this was obviously another of the designs that passed the test of practical service.

Water comes into many of Darwin's other inventions. In his Commonplace Book there is a drawing of a water-closet which is in some respects even more clean and decent than those now in use. Another sketch in the Commonplace Book (p. 83) shows steam playing on a multi-bladed turbine-wheel, Fig. 14, very like the Parsons steam turbine which came into use a century later.

The most complex of Darwin's water machines, shown in *Phytologia* (1800), makes use of a small head of water to raise water to a much greater height. This machine is shown in Fig. 15, and Darwin's description of it is as follows:

> *a b* the stream of water, *b c c* the height of the fall of it, suppose ten feet, *d e* two vessels of lead or iron containing, suppose, four gallons each, *f g h i k l* are vessels of lead containing, suppose, two quarts each, *o p* two cocks, each of which passes through two pipes, opening one and closing the other, *q r* a water balance moving on its centre *s* and turning the two cocks *o* and *p*, alternately, *t u* and *w x* two air-pipes of lead one quarter or half an inch diameter within, *y z*, *y z*, *y z*, water-pipes one inch diameter.
>
> The pipe *b c c* is always full from the stream *a b*; the small cisterns *g i l*, and the large one *d*, are supposed to have been previously full of water, then admit water by turning the cock *o*

through the pipe *c e* into the large cistern *e*. This water will press the air, which was in this cistern *e* up the air-pipe *w x*, and will force the water from the small cisterns *g i l* into the cisterns *h k* and great C. At the same time by opening B, the water and condensed air, which previously existed in the large cistern *d*, and the small ones *f h k*, is discharged at B. After a time the water balance *q r s* closes the cocks now open, and opens their antagonists, and the cisterns *f h k* are emptied in their turn by the force of the condensed air from the cistern *d*, as the water enters into it from the pipe *b c*. (*Phytologia*, Plate VI, after p. 282)

The most massive of Darwin's watery inventions is the canal lift (1777). If a canal has to accomplish a stiff hill-climb, a flight of locks is needed: this delays traffic, may be difficult to build, and allows water to escape from the upper stretches of the canal whenever the flight is used. Darwin saw the merits of the lift, which avoids these troubles, and is especially suitable for linking a canal with a river passing beneath it.

Darwin begins his description of the lift as follows:

> Let a wooden box be constructed so large as to receive a loaded boat. Let the box be join'd to the end of the upper canal and then the boat is admitted, and the doors of admission secured again. Then the box with the boat in it, being balanced on wheels, or levers, is let down, and becomes part of the inferior lock. (Commonplace Book, p. 58)

As shown in Fig. 16, he envisages that the box should hang from chains which run over pulleys and have counterweights to balance the weight of the box. The key to success is, as he notes, the fact that

> nothing is to be overcome but the vis inertia and friction, as the water box is always the same weight, whether the boat is in it or not, since the boat whether full or empty will detrude [displace] just as much water as its own weight.

Darwin at first proposed a metal counterweight, but then he saw the advantages of having a second box identical with the first. This doubles the possible capacity of the lift, and when repairs are needed both caissons can be emptied of water simultaneously without upsetting the balance.

Several ingenious canal lifts were tried out during the 1790s, and the first lift using counterbalanced caissons, as suggested by Darwin, was patented in 1798 and began operating in 1800, at Mells in Somerset. The large Anderton lift near Northwich in Cheshire, opened in 1875, was similar in design. This lift, which still operates in a slightly modified form, raises boats some fifty feet from the Weaver Navigation Canal to the Grand Trunk Canal—the very canal Darwin had been associated with. Many large lifts have since been built in other European countries and the U.S.A.[33]

We have now seen a fair though very small sample of Darwin's inventions. Among the others are several that have been left behind by the advance of technology, such as his improved oil lamps and telescopic candlesticks, his many designs for stocking frames, spinning machines for mohair and various sewing machines. In his more playful moments he dabbled with perpetual motion, designed a machine to move chessmen, and devised a 'factitious spider' which walked round on a salver under the influence of rotating magnets concealed beneath. In more serious moods he designed weighing machines rather like those now seen in chemists' shops; water turbines; surveying machines; the improved drill plough mentioned in Chapter 7; and a continuous-flow rotary pump with toothed wheels, just like the rotary pumps subsequently used.

These recorded inventions are however only a small fraction of the total. Darwin was a compulsive inventor, who jumped from one 'hobby-horse' to another with surprising agility (as shown by his first letter to Watt, in Chapter 2). Most of the inventions sketched in his Commonplace Book belong to the years 1777–9, and we have scarcely any record of his work in the 1760s, probably his most inventive years. Sometimes he invented just for the fun of it; sometimes he was trying to improve a defective existing design; and sometimes he hit on a new idea which could give industry a real stimulus. Whatever the subject, he was always ready with ideas, at Lunar Society meetings or in private conversation; but he was usually quite content to let others take up and develop the invention, and had no thought of claiming priority.

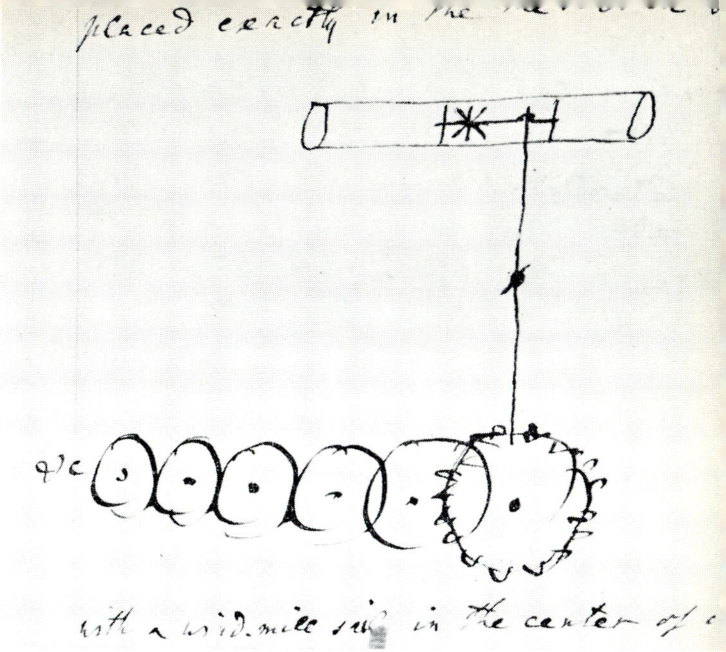

FIG. 9 (*above*). An instrument for measuring the net flow of air from north to south. From Darwin's Commonplace Book, 1777. For explanation, see page 123. FIG. 10 (*below*). The contract signed in 1771 by Boulton, and witnessed by Keir and Small, promising to pay Darwin £1,000 for a developed version of his speaking-machine. See page 125. From Darwin's Commonplace Book.

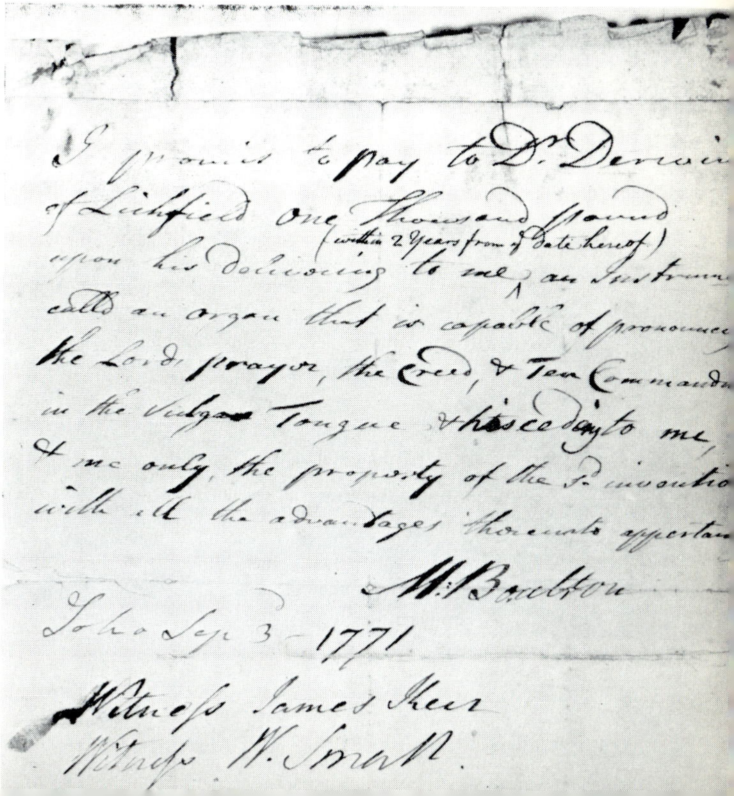

I promise to pay to Dr. Darwi
of Lichfield One Thousand pounds (within 2 Years from ye date hereof)
upon his delivering to me an Instrum
called an organ that is capable of pronounc
the Lords prayer, the Creed, & Ten Commandm
in the Vulgar Tongue & his ceding to me,
& me only, the property of the sd inventio
with all the advantages thereunto appertain

M. Boulton

Soho Sep 3 1771

Witness James Keir
Witness W. Small

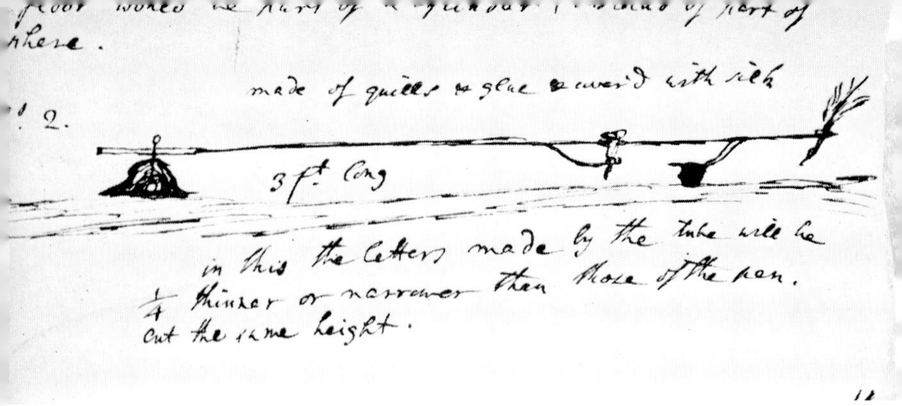

FIG. 11. One of Darwin's 'bigraphers', or copying-machines: the words written by the quill pen on the right are reproduced by the quill in the middle, which has a reservoir of ink above. From Darwin's Commonplace Book, 1777.

FIG. 12. Part of Darwin's design for an artificial bird, from his Commonplace Book, 1777. In this version the model is energized by a wound spring, but Darwin suggests a compressed-air bottle as an alternative source of power. For explanation, see page 125.

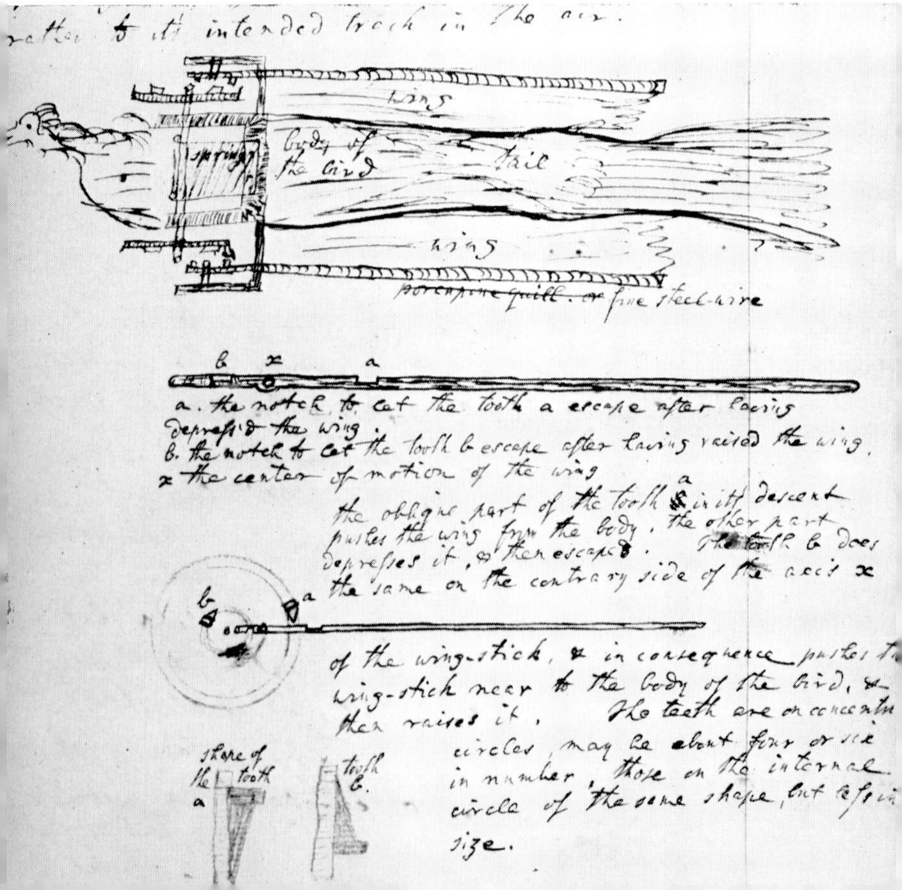

9 Verse in Full Bloom

No writer has surpassed him in the luminous representation of visible objects in verse.

G. L. CRAIK; *History of English Literature*

DARWIN'S first poem, 'The Death of Prince Frederick', was published in 1751, while he was still an undergraduate at Cambridge. But he suppressed his talent for verse to avoid damaging his reputation as a doctor. It was not until he was nearly sixty, in 1789, that he broke his thirty-eight-year silence and burst upon an astonished public with *The Loves of the Plants*, which at once made him the most famous poet of the day. *The Loves of the Plants* was the second and lighter half of a longer poem *The Botanic Garden*, whose publication was completed by the appearance in 1791 of the first and weightier half, entitled *The Economy of Vegetation* (for which Darwin received an advance of one thousand guineas from his publisher). *The Botanic Garden* is the main subject of this chapter.

The thirty-eight fallow years are punctuated by occasional outcrops of verse, usually humorous, like this doggerel he wrote for his children:

O mortal man that liv'st by bread!
What makes thy nose to look so red?
'Tis Burton Ale, so strong and stale,
That keeps my nose from growing pale.

Or these comments on Dr Swan's edition of Sydenham's works, with notes:

The *Swans* of Old had tuneful throats
And utter'd forth melodious *Notes*,
But such the *Notes* our Bird produces
You'd swear the modern Swan a goose is!

(Commonplace Book, p. 165)[34]

In 1789 no one would have guessed, from such squibs as these, that Darwin was about to bring out a poem of 4,384 lines, in technically-perfect rhyming couplets. It was, moreover, a poem that took all scientific knowledge as its subject, and Darwin succeeded in communicating the information in terms that even the literary men like Horace Walpole could understand and applaud. It is an achievement that has never been rivalled.

The Botanic Garden is more than a compendium of science, however: it possesses a lightness of touch which makes it one of the most amusing poems ever written. J. V. Logan, in his book on Darwin's poetry and aesthetics, calls *The Loves of the Plants* 'the culmination of Darwin's humour'.

The first part of *The Botanic Garden*, misleadingly entitled *The Economy of Vegetation*, ranged widely over the realms of science and technology. Its 2,400 lines of verse are supported, and at times overwhelmed, by copious explanatory notes.

The machinery of *The Economy of Vegetation* is quite elaborate—as perhaps befits an author so keen on inventing real machines. Darwin makes the Goddess of Botany descend to Earth, in a 'blushing car' decked with flowers, and well-sprung too:

> And now on earth the silver axle rings,
> And the shell sinks upon its slender springs;
> Light from her airy seat the Goddess bounds,
> And steps celestial press the pansied grounds. (I 65–8)

Around her graceful footsteps gather the Nymphs of Fire and Water, earthy Gnomes, and airy Sylphs, and in the four successive cantos of the poem she addresses these elemental Genii in turn. The exclamations 'Nymphs!' or 'Gnomes!' at the beginning of nearly every paragraph are slightly distracting at first, but a modern reader quickly learns to ignore these pinpricks.

Darwin takes the four elements in the order (1) Fire, (2) Earth, (3) Water and (4) Air. Canto I begins finely with an address to the 'nymphs of primeval fire', describing the origin of the solar system, in which planets issue from the Sun and

> Bend, as they journey with projectile force,
> In bright ellipses their reluctant course. (I 109–10)

Then, with the calm mastery that so impressed his contemporaries, Darwin links the nymphs of fire with accurate descriptions of shooting stars, lightning, rainbows, morning and evening sky colours, an outermost atmosphere of hydrogen, refraction, twilight, fireballs, the aurora, planets, comets, the fixed stars and the Sun. Rarely has so much ground been covered so smartly as in this lightning tour of the heavens:

> Ethereal Powers! you chase the shooting stars,
> Or yoke the vollied lightnings to your cars,
> Cling round the aerial bow with prisms bright,
> And pleased untwist the sevenfold threads of light;
> Eve's silken couch with gorgeous tints adorn,
> And fire the arrowy throne of rising Morn.
> —Or, plum'd with flame, in gay battalions spring
> To brighter regions borne on broader wing;
> Where lighter gases, circumfused on high,
> Form the vast concave of exterior sky;
> With airy lens the scatter'd rays assault,
> And bend the twilight round the dusky vault;
> Ride, with broad eye and scintillating hair,
> The rapid Fire-ball through the midnight air;
> Dart from the North on pale electric streams,
> Fringing Night's sable robe with transient beams.
> —Or rein the Planets in their swift careers,
> Gilding with borrow'd light their twinkling spheres;
> Alarm with comet-blaze the sapphire plain,
> The wan stars glimmering through its silver train;
> Gem the bright Zodiac, stud the glowing pole,
> Or give the Sun's phlogistic orb to roll. (I 115–36)

The fiery catalogue continues with volcanoes and phosphoric lights, glow-worms and gunpowder, to name but a few of the not-so-cold scientific facts. To lighten the poem even more, Darwin adds episodes from classical mythology: these are usually irreverent, sometimes irrelevant, quite often erotic, and nearly always amusing. The poem does need lightening at times, for Darwin likes to describe familiar actions in ponderous scientific terms, and it was this feature of his style that provoked parody later in the 1790s. For example, in the firing of a cannon it is not just a matter of 'light the touch paper and stand clear'; instead he offers us full details—

Pent in dark chambers of cylindric brass,
Slumbers in grim repose the sooty mass;
Lit by the brilliant spark, from grain to grain
Runs the quick fire along the kindling train;
On the pain'd ear-drum bursts the sudden crash,
Starts the red flame, and Death pursues the flash. (I 243–8)

Before this resounding report has died away, steam engines appear on the scene. Darwin mentions all their uses, and includes an account of Boulton's coining machines:

With iron lips his rapid rollers seize
The lengthening bars, in thin expansion squeeze;
Descending screws with ponderous fly-wheels wound
The tawny plates, the new medallions round;
Hard dyes of steel the cupreous circles cramp,
And with quick fall his massy hammers stamp.
The Harp, the Lily and the Lion join,
And GEORGE and BRITAIN guard the sterling coin. (I 281–8)

Darwin believed it would be possible to build flying-machines propelled 'by the power of steam, or some other explosive material, which another half century may probably discover'. This belief pervades the famous couplets in which he foresees the use of aircraft in war:

Soon shall thy arm, UNCONQUER'D STEAM! afar
Drag the slow barge, or drive the rapid car;
Or on wide-waving wings expanded bear
The flying-chariot through the fields of air.
—Fair crews triumphant, leaning from above,
Shall wave their fluttering kerchiefs as they move;
Or warrior-bands alarm the gaping crowd,
And armies shrink beneath the shadowy cloud. (I 289–96)

Electrical machines come next, conducting us to a lovers' meeting where the sparks really fly:

Or, if on wax some fearless Beauty stand,
And touch the sparkling rod with graceful hand;
Through her fine limbs the mimic lightnings dart,

And flames innocuous eddy round her heart;
O'er her fair brow the kindling lustres glare,
Blue rays diverging from her bristling hair;
While some fond Youth the kiss ethereal sips,
And soft fires issue from their meeting lips. (I 349–56)

The remainder of the canto is also quite electrifying. Finally the Nymphs of Fire depart in a firework display:

Red rockets rise, loud cracks are heard on high,
And showers of stars rush headlong from the sky,
Burst, as in silver lines they hiss along,
And the quick flash unfolds the gazing throng. (I 595–8)

The second canto is addressed to the Gnomes of Earth, and begins with another picture of the Earth's creation:

Gnomes! your bright forms, presiding at her birth,
Clung in fond squadrons round the new-born EARTH;
When high in ether, with explosion dire,
From the deep craters of his realms of fire,
The whirling Sun this ponderous planet hurl'd,
And gave the astonish'd void another world.
When from its vaporous air, condensed by cold,
Descending torrents into oceans roll'd;
And fierce attraction with relentless force
Bent the reluctant wanderer to its course. (II 11–20)

Earth's course was far from smooth, however: it suffered a nasty shock, when the Moon was born.

Gnomes! how you shriek'd! when through the troubled air
Roar'd the fierce din of elemental war;
When rose the continents, and sunk the main,
And Earth's huge sphere exploding burst in twain.—
Gnomes! how you gazed! when from her wounded side
Where now the South-Sea heaves its waste of tide,
Rose on swift wheels the MOON's refulgent car,
Circling the solar orb, a sister-star,
Dimpled with vales, with shining hills emboss'd,
And roll'd round Earth her airless realms of frost. (II 73–82)

Darwin's speculation that the Moon came out of what is now the Pacific was given mathematical form a hundred years later by his great-grandson Sir George Darwin. Subsequently the theory has been rejected, and revived: the question of the Moon's origin is still undecided.

Darwin next explains how limestone forms from shells, and how morasses ooze out salts of iron. This leads to praise of 'adamantine steel', the basis for the compass, 'the plowshare and the sword'. Then Darwin discusses acids, precious stones, and clays, which in turn lead inevitably to pottery and to Wedgwood:

> And pleased on WEDGWOOD ray your partial smile,
> A new Etruria decks Britannia's isle. . . .
> Charmed by your touch, the kneaded clay refines,
> The biscuit hardens, the enamel shines;
> Each nicer mould a softer feature drinks,
> The bold Cameo speaks, the soft Intaglio thinks. . . .
>
> Whether, O Friend of Art! your gems derive
> Fine forms from Greece, and fabled Gods revive;
> Or bid from modern life the Portrait breathe,
> And bind round Honour's brow the laurel wreath;
> Buoyant shall sail, with Fame's historic page,
> Each fair medallion o'er the wrecks of age.
>
> (II 303–4, 307–10, 341–6)

Flashing past coal, jet and amber, we come, via electricity, Franklin, America and liberty, to the French Revolution:

> Long had the Giant-form [Liberty] on GALLIA's plains
> Inglorious slept, unconscious of his chains;
> Round his large limbs were wound a thousand strings
> By the weak hands of Confessors and Kings;
> O'er his closed eyes a triple veil was bound,
> And steely rivets lock'd him to the ground;
> While stern Bastille with iron cage inthralls
> His folded limbs, and hems in marble walls.
> —Touch'd by the patriot-flame, he rent amazed
> The flimsy bonds, and round and round him gazed.
>
> (II 377–386)

Tin, copper, zinc and lead are awarded only one line each, but the Spanish conquest of Mexico and Peru for the sake of their mineral wealth makes Darwin furious:

Heavens! on my sight what sanguine colours blaze!
Spain's deathless shame! the crimes of modern days!
When Avarice, shrouded in Religion's robe,
Sail'd to the West, and slaughter'd half the globe;
While Superstition, stalking by his side,
Mock'd the loud groans, and lapp'd the bloody tide.
(II 413–18)

Britain too is pummelled for still allowing the slave-trade:

Hear, oh, BRITANNIA! potent Queen of isles,
On whom fair Art, and meek Religion smiles,
How AFRIC's coasts thy craftier sons invade
With murder, rapine, theft,—and call it Trade!
—The SLAVE, in chains, on supplicating knee,
Spreads his wide arms, and lifts his eyes to Thee;
With hunger pale, with wounds and toil oppress'd,
'ARE WE NOT BRETHREN?' sorrow choaks the rest;—
—AIR! bear to heaven upon thy azure flood
Their innocent cries!—EARTH! cover not their blood!
(II 421–30)

Darwin was so keen to see 'blood-nursed Tyrants' receive their just deserts that he is lured into a long and racy digression about a military adventure of Cambyses which ran into the sands: for when he sent his army across the desert, the Gnomes of Earth buried it in sand. Here is the climax—a stifling end to a vigorous tale:

Onward resistless rolls the infuriate surge,
Clouds follow clouds, and mountains mountains urge;
Wave over wave the driving desert swims,
Bursts o'er their heads, inhumes their struggling limbs;
Man mounts on man, on camels camels rush,
Hosts march o'er hosts, and nations nations crush,—
Wheeling in air the winged islands fall,
And one great earthy Ocean covers all! (II 487–94)

The Gnomes are also given credit for making things grow—they are the genii of the good earth, as well as the sand-devils. The canto ends as they trip away in 'dusky squadrons and in shining groups'.

The third canto introduces the 'aquatic Nymphs', who

lead with viewless march
The wingèd vapours up the aerial arch,

—and then lead them down again as rivers, to the sea, an endless cycle as essential as that of blood circulation in animals. The nymphs ride the tides and play in fishy realms:

You chase the warrior Shark, and cumbrous Whale,
And guard the Mermaid in her briny vale;
Feed the live petals of her insect-flowers,
Her shell-wrack gardens, and her sea-fan bowers.
(III 79–82)

Again Darwin's fascination with coral emerges:

You stay the flying bark, conceal'd beneath,
Where living rocks of worm-built coral breathe.
(III 89–90)

The theme of 'water, water everywhere' continues with rivers from beneath Alpine snows, tropical monsoons, Icelandic geysers, the warm springs of Buxton, and the gardens of Chatsworth, where the water-nymphs disport themselves:

Oft by her bubbling founts, and shadowy domes,
In gay undress the fairy legion roams,
Their dripping palms in playful malice fill,
Or taste with ruby lip the sparkling rill. (III 175–8)

The Nymphs are also well versed in the new French chemistry, and they recognize the powerful chemical bond between oxygen (pure Air) and hydrogen (flaming Gas), which makes H_2O such a stable compound:

Nymphs! your bright squadrons watch with chemic eyes
The cold-elastic vapours, as they rise;
With playful force arrest them as they pass,
And to *pure* AIR betroth the *flaming* GAS.
Round their translucent forms at once they fling
Their rapturous arms, with silver bosoms cling.
(III 201–6)

The effect is ludicrous: but of course that is just what Darwin intended, for he is at his favourite game of amusing the reader.

Water's power to beautify a landscape evokes a delicate tribute:

Nymphs of aquatic Taste! whose placid smile
Breathes sweet enchantment o'er BRITANNIA's isle. . . .
You with nice ear, in tiptoe trains, pervade
Dim walks of morn or evening's silent shade;
Join the lone Nightingale, her woods among,
And roll your rills symphonious to her song.
(III 261–2, 267–70)

From beauty he turns to utility, with canals and a tribute to Brindley:

Your virgin trains on BRINDLEY's cradle smiled,
And nursed with fairy-love the unletter'd child,
Spread round his pillow all your secret spells,
Pierced all your springs, and open'd all your wells. . . .
So with strong arm immortal BRINDLEY leads
His long canals, and parts the velvet meads;
Winding in lucid lines, the watery mass
Mines the firm rock, or loads the deep morass. . . .
(III 321–4, 329–32)

And so to Darwin's description of water pumps—

Nymphs! you first taught to pierce the secret caves
Of humid earth, and lift her ponderous waves;
Bade with quick stroke the sliding piston bear
The viewless columns of incumbent air;—
Press'd by the incumbent air the floods below,
Through opening valves in foaming torrents flow.
(III 345–50)

This leads to pumps for putting out fires and the draining of morasses. After another brief display of their rain-making powers, the Nymphs move off:

Then with soft murmur sweep in lucid trains
Down the green slopes, and o'er the pebbly plains,
To each bright stream on silver sandals glide,
Reflective fountain, and tumultuous tide.
(III 551–4)

In the fourth and final canto of *The Economy of Vegetation* we meet the Sylphs of Air, who make the breezes blow and plants exude oxygen. In sourer moods they bring fogs, the sirocco, simoom and tornado. As patron saints of atmospheric science, the Sylphs taught Toricelli and Boyle

The spring and pressure of the viewless air.
How up exhausted tubes bright currents flow
Of liquid silver from the lake below,
Weigh the long column of the incumbent skies,
And with the changeful moment fall and rise.
(IV 130–4)

They also inspired Mongolfier, the pioneer of ballooning:

Sylphs! your soft voices, whispering from the skies,
Bade from low earth the bold MONGOLFIER rise;
Outstretch'd his buoyant ball with airy spring,
And bore the Sage on levity of wing. (IV 143–6)

The discovery of oxygen by the 'sage' Priestley suggested to Darwin that long journeys could be made by submarine:

Led by the Sage, Lo! Britain's sons shall guide
Huge SEA-BALLOONS beneath the tossing tide;
The diving castles, roof'd with spheric glass,
Ribb'd with strong oak, and barr'd with bolts of brass,
Buoy'd with pure air shall endless tracks pursue,
And PRIESTLEY's hand the vital flood renew. . . .
Deep, in warm waves beneath the Line that roll,
Beneath the shadowy ice-isles of the Pole . . .
(IV 195–200, 205–6)

Darwin wishes the Sylphs would explain how to control the winds:

Oh, SYLPHS! disclose in this inquiring age
One GOLDEN SECRET to some favour'd sage;
Grant the charm'd talisman, the chain, that binds,
Or guides the changeful pinions of the wind!
(IV 307–10)

There follows a nicely-phrased but slightly irrelevant reference to Herschel's theory that the stars would eventually rush together and coalesce:

Roll on, YE STARS! exult in youthful prime,
Mark with bright curves the printless steps of Time;
Near and more near your beamy cars approach,
And lessening orbs on lessening orbs encroach;—
Flowers of the sky! ye too to age must yield,
Frail as your silken sisters of the field! (IV 367–72)

Finally, as if to prove that the poem's title, *The Economy of Vegetation*, is not just another of his jokes, Darwin begins to discuss vegetation economically, cramming much of the subject matter of *Phytologia* into the last 200 lines of the poem. At the end the Goddess of Botany calls the Sylphs to

Bring my rich Balms from Mecca's hallow'd glades,
Sweet flowers, that glitter in Arabia's shades;
Fruits, whose fair forms in bright succession glow
Gilding the Banks of Arno, or of Po;
Each leaf, whose fragrant steam with ruby lip
Gay China's nymphs from pictur'd vases sip;
Each spicy rind, which sultry India boasts,
Scenting the night-air round her breezy coasts.
(IV 595–602)

Loaded with these delicacies, the Goddess flies off into the upper air.

Part II of *The Botanic Garden*, entitled *The Loves of the Plants*, really does live up to its title, by detailing and humanizing the methods of fertilization in flowers. This second part of the poem was published first, in 1789, and won Darwin great acclaim. The versification of science was minute and skilful; and the humanization of the pistils and stamens of the plants was thought slightly scandalous. Today, however, these two innovations have lost their appeal, and the repeated descriptions of vegetable courtship are somewhat tedious. Darwin recognizes this danger and lays on an entertaining programme of diversions, which he calls the Poetic Exhibition—it includes such varied items as 'Parched deserts in Africa', 'Turkish lady in an undress', 'Group of wild creatures drinking' and 'Witch and imps in a church'.

Darwin begins by explaining, with illustrations, the 24 Linnaean classes of plants, distinguished by the characters of the males (stamens) in each flower: Class 1, *Monandria*, has one stamen per flower; Class 2, *Diandria*, has two stamens; and so on, down to Class 13, *Polyandria*, many males. After that come more complex systems, such as Class 15, *Tetradynamia*, 'four powers', with six stamens, four being taller than the other two. Darwin gives examples of each of these types in turn, and Canto I in particular is sometimes like a catalogue.

It all sounds very didactic; to make a best-seller out of such unpromising material was an amazing achievement. Darwin may have been ponderous and clumsy in real life, but in verse his touch was remarkably delicate, and the poem is continuously amusing. Some critics have been idiotic enough to think the humour was unintentional. But Darwin's prime aim was to avoid solemnity, as his proem indicates:

> Gentle reader!
> Lo, here a CAMERA OBSCURA is presented to thy view, in which are lights and shades dancing on a whited canvas, and magnified into apparent life!—if thou art perfectly at leisure for such trivial amusement, walk in, and view the wonders of my INCHANTED GARDEN.
>
> Whereas P. OVIDIUS NASO, a great Necromancer in the famous Court of AUGUSTUS CAESAR, did by art poetic transmute Men, Women, and even Gods and Goddesses, into Trees and Flowers; I have undertaken by similar art to restore some of them to their original animality, after having remained prisoners so long in their respective vegetable mansions; and have here exhibited them before thee. Which thou may'st contemplate as diverse little pictures suspended over the chimney of a Lady's dressing-room, *connected only by a slight festoon of ribbons*. And which, though thou may'st not be acquainted with the originals, may amuse thee by the beauty of their persons, their graceful attitudes, or the brilliancy of their dress.
>
> FAREWELL

Darwin begins the poem by calling on the Sylphs and Gnomes to listen

> While in soft notes I tune to oaten reed
> Gay hopes, and amorous sorrows of the mead.—
> From giant Oaks, that wave their branches dark,
> To the dwarf Moss, that clings upon their bark.

What Beaux and Beauties crowd the gaudy groves,
And woo and win their vegetable Loves.
How Snowdrops cold, and blue-eyed Harebels blend
Their tender tears, as o'er the stream they bend;
The lovesick Violet, and the Primrose pale
Bow their sweet heads, and whisper to the gale;
With secret sighs the Virgin Lily droops,
And jealous Cowslips hang their tawny cups.
How the young Rose in beauty's damask pride
Drinks the warm blushes of his bashful bride;
With honey'd lips enamour'd Woodbines meet,
Clasp with fond arms, and mix their kisses sweet.

(I 5–20)

Darwin would be amused to know that his joke about the Woodbines has since flowered so pungently.

Darwin then calls on the Botanic Muse to

Say on each leaf how tiny Graces dwell;
How laugh the Pleasures in a blossom's bell;
How insect Loves arise on cobweb wings,
Aim their light shafts, and point their little stings.

(I 35–8)

First we have the virtuous *Canna*, Indian reed, with one male and one female in each flower 'plighting their nuptial vow'. Next come the less proper *Callitrache* (stargrass), where two virgins share one male, and *Collinsonia*, where two swains share one female. With *Genista* (Dyer's broom), *Melissa* and *Maedia*, the polygamy is getting out of hand:

Sweet blooms GENISTA in the myrtle shade,
And *ten* fond brothers woo the haughty maid.
Two knights before thy fragrant altar bend,
Adored MELISSA! and *two* squires attend.
MAEDIA's soft chains *five* suppliant beaux confess,
And hand in hand the laughing belle address;
Alike to all, she bows with wanton air,
Rolls her dark eye, and waves her golden hair.

(I 57–64)

Melissa has four stamens, two standing higher than the others: hence the two *knights* and two *squires*. Darwin may be skittish, but he is also exact.

So the catalogue continues, at a 'light and skipping pace',[35] with many comic touches. In *Lychnis*, for example, the five females 'rise above the petals, as if looking abroad for their distant husbands':

> Each wanton beauty, trick'd in all her grace,
> Shakes the bright dew-drops from her blushing face;
> In gay undress displays her rival charms,
> And calls her wondering lovers to her arms. (I 113–16)

Darwin soon brings in his favourite insectivorous plants, first *Silene* (catchfly), which has a sticky substance to trap insects:

> Haste, glittering nations, tenants of the air,
> Oh, steer from hence your viewless course afar!
> If with soft words, sweet blushes, nods, and smiles,
> The *three* dread Syrens lure you to their toils,
> Limed by their art in vain you point your stings,
> In vain the efforts of your whirring wings! (I 143–8)

Then there is *Drosera*, sun-dew, later investigated in great detail by Charles Darwin:

> Queen of the marsh, imperial DROSERA treads
> Rush-fringed banks, and moss-embroider'd beds;
> Redundant folds of glossy silk surround
> Her slender waist, and trail upon the ground. (I 231–4)

Another of Darwin's favourites is *Mimosa*, the sensitive plant:

> Weak with nice sense, the chaste MIMOSA stands,
> From each rude touch withdraws her timid hands. . . .
> Shuts her sweet eye-lids to approaching night,
> And hails with freshen'd charms the rising light.
> Veil'd, with gay decency and modest pride,
> Slow to the mosque she moves, an eastern bride.
> (I 301–2, 307–10)

The canto ends at evening in lines reminiscent of Gray, Collins and Pope:

> Now the light swallow with her airy brood
> Skims the green meadow, and the dimpled flood;

Loud shrieks the lone thrush from his leafless thorn,
Th' alarmèd beetle sounds his bugle horn;
Each pendent spider winds with fingers fine
His ravell'd clue, and climbs along the line;
Gay Gnomes in glittering circles stand aloof
Beneath a spreading mushroom's fretted roof;
Swift bees returning seek their waxen cells,
And Sylphs cling quivering in the lily's bells.
Through the still air descend the genial showers,
And pearly rain-drops deck the laughing flowers. (I 473–84)

Before embarking on Canto II, we enjoy an Interlude, filled by conversation between Darwin and his Bookseller (i.e. publisher):

Bookseller. Your verses, Mr. Botanist, consist of *pure description*, I hope there is *sense* in the notes.

Poet. I am only a flower-painter, or occasionally attempt a landskip; and leave the human figure with the subjects of history to abler artists. . . .

Bookseller. . . . pray tell me, what is the essential difference between Poetry and Prose? . . .

Poet. Next to the measure of the language, the principal distinction appears to me to consist in this: that Poetry admits of but few words expressive of very abstracted ideas, whereas Prose abounds with them. And as our ideas derived from visible objects are more distinct than those derived from the objects of our other senses, the words expressive of these ideas belonging to vision make up the principal part of poetic language. That is, the Poet writes principally to the eye, the Prose-writer uses more abstracted terms. (pp. 47–8)

Later in the dialogue Darwin draws the further distinction that prose often instructs, while poetry 'amuses'. This shows that Darwin's aim as a poet was to create amusing pictures.

In the second canto the catalogue of fertilization rites continues, but with more and livelier digressions. The flying seeds of the thistle lead on to ballooning:

So on the shoreless air the intrepid Gaul
Launch'd the vast concave of his buoyant ball.
Journeying on high, the silken castle glides
Bright as a meteor through the azure tides. . . .

The calm Philosopher in ether sails,
Views broader stars,* and breathes in purer gales;
Sees, like a map, in many a waving line
Round Earth's blue plains her lucid waters shine;
Sees at his feet the forky lightnings glow,
And hears innocuous thunders roar below.
—Rise, great MONGOLFIER! urge thy venturous flight
High o'er the Moon's pale ice-reflected light;
High o'er the pearly Star, whose beamy horn
Hangs in the east, gay harbinger of morn. (II 25–8, 41–50)

Gossypium, the cotton plant, provokes Darwin to describe spinning machines in great detail:

First with nice eye emerging Naiads cull
From leathery pods the vegetable wool;
With wiry teeth *revolving cards* release
The tangled knots, and smooth the ravell'd fleece;
Next moves the *iron-hand* with fingers fine,
Combs the wide card, and forms the eternal line;
Slow, with soft lips, the *whirling Can* acquires
The tender skeins, and wraps in rising spires;
With quicken'd pace *successive rollers* move,
And these retain, and those extend the *rove*;
Then fly the spoles, the rapid axles glow,
And slowly circumvolves the labouring wheel below.
(II 93–104)

This was one of the passages parodied later in the 1790s, in *The Loves of the Triangles*.

When Darwin reaches *Menispermum*, the Indian berry, which is said to intoxicate fish, he seizes the chance to present a ludicrous picture of St. Anthony of Padua preaching to the fishes:

So, when the Saint from Padua's graceless land
In silent anguish sought the barren strand,
High on the shatter'd beech sublime He stood,
Still'd with his waving arm the babbling flood;
'To Man's dull ear,' He cry'd, 'I call in vain,

* I.e. a broader expanse of stars. This is one of Darwin's rare lapses in lucidity.

'Hear me, ye scaly tenants of the main!'—
Misshapen Seals approach in circling flocks,
In dusky mail the Tortoise climbs the rocks,
Torpedoes, Sharks, Rays, Porpus, Dolphins, pour
Their twinkling squadrons round the glittering shore;
With tangled fins, behind, huge Phocae glide,
And Whales and Grampi swell the distant tide.
Then kneel'd the hoary Seer, to heaven address'd
His fiery eyes, and smote his sounding breast;
'Bless ye the Lord', with thundering voice he cry'd,
'Bless ye the Lord!' the bending shores reply'd;
The winds and waters caught the sacred word,
And mingling echoes shouted 'Bless the Lord!'
The listening shoals the quick contagion feel,
Pant on the floods, inebriate with their zeal,
Ope their wide jaws, and bow their slimy heads,
And dash with frantic fins their foamy beds. (II 245–66)

The sidelong slaps at religion which marked Darwin's conversation seem here to have slipped into his verse.

After paying tribute to the Peruvian bark and foxglove for their powers of restoring health, Darwin goes on to discuss man's well-being in general, and Howard's philanthropy in particular:

From realm to realm, with cross or crescent crown'd,
Where'er Mankind and Misery are found,
O'er burning sands, deep waves, or wilds of snow,
Thy HOWARD journeying seeks the house of woe.
Down many a winding step to dungeons dank,
Where anguish wails aloud, and fetters clank;
To caves bestrew'd with many a mouldering bone,
And cells, whose echoes only learn to groan;
Where no kind bars a whispering friend disclose,
No sunbeam enters, and no zephyr blows,
He treads, inemulous of fame or wealth,
Profuse of toil, and prodigal of health. (II 443–54)

The Botanic Muse then pauses for a well-earned tea-break:

And now her vase a modest Naiad fills
With liquid crystal from her pebbly rills;

Piles the dry cedar round her silver urn,
(Bright climbs the blaze, the crackling faggots burn),
Culls the green herb of China's envy'd bowers,
In gaudy cups the steamy treasure pours;
And, sweetly smiling, on her bended knee
Presents the fragrant quintessence of Tea. (II 477–84)

Refreshed by this decoction, the Goddess calls her audience to order for the third Canto:

And now the Goddess sounds her silver shell,
And shakes with deeper tones the inchanted dell.
(III 1–2)

These lines warn us that the third Canto will be more sombre than its predecessors, with nightshade, laurel and the upas tree among the *dramatis personae*. In describing nightshade, 'two males, one female', Darwin takes the chance to exploit the 'Gothic' idiom which was coming into fashion—

Pale shoot the stars across the troubled night,
The tim'rous moon withholds her conscious light;
Shrill scream the famish'd bats, and shivering owls,
And loud and long the dog of midnight howls!—
Then yawns the bursting ground!—*two* imps obscene
Rise on broad wings, and hail the baleful queen;
Each with dire grin salutes the potent wand,
And leads the Sorceress with his sooty hand;
Onward they glide, where sheds the sickly yew
O'er many a mouldering bone its nightly dew.
(III 11–20)

Darwin had Gothic scenery close at hand, in Dovedale, and he describes a cave near Wetton:

Rocks rear'd on rocks in huge disjointed piles
Form the tall turrets, and the lengthen'd ailes;
Broad ponderous piers sustain the roof, and wide
Branch the vast rain-bow ribs from side to side.
While from above descends in milky streams
One scanty pencil of illusive beams,
Suspended crags and gaping gulphs illumes,
And gilds the horrors of the deepen'd glooms.
(III 92–100)

As we might expect, the vine winds its way into this sombre canto, with the female luring five males to alcoholic stupor:

> *Five* hapless swains with soft assuasive smiles
> The harlot meshes in her deathful toils;
> 'Drink deep,' she carols, as she waves in air
> The mantling goblet, 'and forget your care'. . . .
> Fell Gout peeps grinning through the flimsy scene,
> And bloated Dropsy pants behind unseen.
> (III 361–4, 367–8)

Darwin ends the canto with another spirited attack on slavery:

> E'en now in Afric's groves with hideous yell
> Fierce SLAVERY stalks, and slips the dogs of hell;
> From vale to vale the gathering cries rebound,
> And sable nations tremble at the sound!—
> —YE BANDS OF SENATORS! whose suffrage sways
> Britannia's realms, whom either Ind obeys;
> Who right the injured, and reward the brave,
> Stretch your strong arm, for ye have power to save!
> . . . hear this truth sublime,
> 'HE, WHO ALLOWS OPPRESSION, SHARES THE CRIME'.
> (III 441–8, 455–6)

In the final canto of *The Loves of the Plants*, the Goddess 'strings her lute to sweeter tones' and turns to happier scenes. Fields of oats beside the lawns of Tweed remind Darwin that oaten straws were the first musical instruments. So he gives us a song, not unlike that in Milton's *Comus*:

> Sweet ECHO! sleeps thy vocal shell,
> Where this high arch o'erhangs the dell;
> While Tweed with sun-reflecting streams
> Chequers thy rocks with dancing beams?—
>
> Here may no clamours harsh intrude,
> No brawling hound or clarion rude;
> Here no fell beast of midnight prowl,
> And teach thy tortured cliffs to howl! (IV 79–86)

And all nature listens:

From ozier bowers the brooding Halcyons peep,
The Swans pursuing cleave the glassy deep,
On hovering wings the wondering Reed-larks play,
And silent Bitterns listen to the lay. (IV 95–8)

This is followed by a pleasant description of Snowdrop,

Warm with sweet blushes bright GALANTHA glows,
And prints with frolic step the melting snows:
O'er silent floods, white hills, and glittering meads
Six rival swains the playful beauty leads,
Chides with her dulcet voice the tardy Spring,
Bids slumbering Zephyr stretch his folded wing,
Wakes the hoarse Cuckoo in his gloomy cave,
And calls the wondering Dormouse from his grave,
Bids the mute Redbreast cheer the budding grove,
And plaintive Ringdove tune her notes to love.
(IV 137–46)

Trapa, which grows partly above and partly below water, rouses her four male admirers to a fine frenzy:

Now all as one they rise with frolic spring,
And beat the wondering air on humid wing;
Now all descending plunge beneath the main,
And lash the foam with undulating train;
Above, below, they wheel, retreat, advance,
In air and ocean weave the mazy dance;
Bow their quick heads, and point their diamond eyes,
And twinkle to the sun with ever-changing dyes.
(IV 213–20)

In its comic intent and fishy flavour this dance resembles Lewis Carroll's lobster quadrille.

Darwin does not forget the lichens growing in the far north, where

Slow o'er the printed snows with silent walk
Huge shaggy forms across the twilight stalk. (IV 349–50)

And he produces a comic picture of the truffle, a fungus that never appears above ground:

Deep, in wide caverns and their shadowy ailes,
Daughter of Earth, the chaste TRUFFELIA smiles;

On silvery beds, of soft asbestus wove,
Meets her Gnome-husband, and avows her love.
(IV 395–8)

The wild fig is fertilized by a gnat which carries pollen from stamen to pistil:

Quick, the light Gnat her airy Lord bestrides,
With cobweb reins the flying courser guides,
From crystal steeps of viewless ether springs,
Cleaves the soft air on still expanded wings;
Darts like a sunbeam o'er the boundless wave,
And seeks the beauty in her *secret* cave. (IV 423–8)

As a fitting finale to all the fecundity we have *Adonis*, with many males and females living together in the same flower:

A *hundred* virgins join a *hundred* swains,
And fond ADONIS leads the sprightly trains;
Pair after pair, along his sacred groves
To Hymen's fane the bright procession moves;
Each smiling youth a myrtle garland shades,
And wreaths of roses veil the blushing maids;
Light Joys on twinkling feet attend the throng,
Weave the gay dance, or raise the frolic song.
(IV 471–8)

This provokes a comparison with the society of the Areoi in Tahiti, 'about 100 males and 100 females, who form one promiscuous marriage':

Thus where pleased VENUS, in the southern main,
Sheds all her smiles on Otaheite's plain,
Wide o'er the isle her silken net she draws,
And the Loves laugh at all but Nature's laws.
(IV 487–90)

This happiest of curtain-lines almost justifies an ambiguous amendment of Pope's famous couplet, to:

Nature and Nature's laws lay hid in night
Till Darwin came and put the matter right.

10 The Ascent of Man

Observe Humanity's advance,
And Evolution's giant strides.
MAY KENDALL; *Taking Long Views*

DARWIN'S second long poem, *The Temple of Nature*, or *The Origin of Society*, was published after his death, in 1803. It attracted much less attention than *The Botanic Garden*, because the climate of opinion had changed so much: in the 1780s the questioning of accepted ideas was at least allowed, and sometimes even welcomed; but after 1792 the atmosphere became repressive, because the French Revolution and the Napoleonic war made the ruling classes nervous. In *The Temple of Nature* Darwin expresses his belief that life originated as microscopic specks in primeval seas and gradually, under evolutionary pressures and without assistance from any deity, developed its present forms. Even readers who had delighted in *The Botanic Garden* did not want their accepted ideas undermined, especially at a time when they were relying on the Deity to help in warding off the arch-fiend Napoleon, who was poised for a landing on the South Coast of England. 'The times were out of joint' for Darwin's poem; but that does nothing to detract from his remarkable achievement in versifying evolution so completely, over fifty years before it began to be accepted by scientists. Even without this prophetic insight, *The Temple of Nature* would be a better poem than that encyclopaedic mixture *The Botanic Garden*: it has a sound structure, and moves forward logically.

Darwin explains his intentions in the Preface:

> The Poem, which is here offered to the Public, does not pretend to instruct by deep researches of reasoning; its aim is simply to amuse by bringing distinctly to the imagination the beautiful and sublime images of the operations of Nature in the order, as the Author believes, in which the progressive course of time presented them.

Again we need to underline and remember the words *its aim*

is simply to amuse, and be on the look-out for the ludicrous.

Like its predecessor, *The Temple of Nature* is helped along by machinery:

> In the Eleusinian mysteries the philosophy of the works of Nature, with the origin and progress of society, are believed to have been taught by allegoric scenery explained by the Hierophant to the initiated, which gave rise to the machinery of the following Poem. (From Preface)

This explains the poem's title: most of the story is told by the Priestess of Nature, Urania, who resides in the Temple of Nature.

In Canto I the machinery seems at first to have taken charge of the poem completely. We are offered an impressive picture of the Temple of Nature:

> Here, high in air, unconscious of the storm,
> Thy temple, NATURE, rears its mystic form;
> From earth to heav'n, unwrought by mortal toil,
> Towers the vast fabric on the desert soil;
> O'er many a league the ponderous domes extend,
> And deep in earth the ribbed vaults descend;
> A thousand jasper steps with circling sweep
> Lead the slow votary up the winding steep;
> Ten thousand piers, now join'd and now aloof,
> Bear on their branching arms the fretted roof. (I 65–74)

After going down the spiral staircase of a thousand jasper steps, the votaries may be forgiven for becoming giddy:

> Here o'er piazza'd courts, and long arcades,
> The bowers of PLEASURE root their waving shades;
> Shed o'er the pansied moss a checker'd gloom,
> Bend with new fruits, with flow'rs successive bloom.
> Pleas'd, their light limbs on beds of roses press'd,
> In slight undress recumbent Beauties rest;
> On tiptoe steps surrounding Graces move,
> And gay Desires expand their wings above. (I 89–96)

The prospect of Beauties reclining on a rose-bed in a state of 'slight undress' inspired a more recent poet to record the fate of shapely Rose:

As shapely Rose
lay on the lawn
in classic pose,
A shapely rose
sporting a thorn
impaled her toes.
Shapely Rose sharply rose.

Darwin spices *The Temple of Nature* with many amusing episodes, from the nice ambiguity of

On sun-bright lawns unclad the Graces stray'd,
And guiltless Cupids haunted every glade,

to the mock-heroic,

Now rocks on rocks, in savage grandeur roll'd,
Steep above steep, the blasted plains infold. (I 39–40, 47–8)

And in some of his routine descriptions the couplets are as neat and polished as any of Pope's:

Now rose in purple pomp the breezy dawn,
And crimson dew-drops trembled on the lawn;
Blaz'd high in air the temple's golden vanes,
And dancing shadows veer'd upon the plains.—
Long trains of virgins from the sacred grove,
Pair after pair, in bright procession move,
With flower-fill'd baskets round the altar throng,
Or swing their censers, as they wind along. (I 155–62)

However, these pleasantries must not divert us any longer: the main theme of Canto I is the slow evolution of microscopic life in warm primeval oceans, with the subsequent gradual transfer to an amphibious existence, and finally to life on land. The evolutionary sections of *The Temple of Nature* have already been adequately sampled in Chapter 6: the quotations to be given now serve rather to indicate Darwin's general picture of life. Evolution still finds its way in at times, however, as in the following summary of childbirth, in which the progress of the embryo from water to land is tacitly compared with evolutionary development:

Thus in the womb the nascent infant laves
Its natant form in the circumfluent waves;
With perforated heart unbreathing swims,
Awakes and stretches all its recent limbs;
With gills placental seeks the arterial flood,
And drinks pure ether from its Mother's blood.
Erewhile the landed stranger bursts his way,
From the warm wave emerging into day;
Feels the chill blast, and piercing light, and tries
His tender lungs, and rolls his dazzled eyes;
Gives to the passing gale his curling hair,
And steps a dry inhabitant of air. (I 389–400)

This prepares us for Canto II, 'On the Reproduction of Life':

How short the span of LIFE! some hours possess'd,
Warm but to cool, and active but to rest!—
The age-worn fibres goaded to contract,
By repetition palsied, cease to act. . . .
But REPRODUCTION with ethereal fires
New life rekindles, ere the first expires. . . .
Organic forms with chemic changes strive,
Live but to die, and die but to revive!
Immortal matter braves the transient storm,
Mounts from the wreck, unchanging but in form.
(II 1–4, 13–14, 41–4)

Asexual reproduction came first, Darwin believes:

The Reproductions of the living Ens
From sires to sons, unknown to sex, commence. . . .
So the male Polypus parental swims,
And branching infants bristle all his limbs;
So the lone Taenia,* as he grows, prolongs
His flatten'd form with young adherent throngs;
Unknown to sex the pregnant oyster swells;
And coral-insects build their radiate shells. . . .
Birth after birth the line unchanging runs,
And fathers live transmitted in their sons;
Each passing year beholds the unvarying kinds,

* Tape-worm.

The same their manners, and the same their minds.
(II 63–4, 85–90, 107–10)

The defects of asexual reproduction, already expounded in *Phytologia*, are neatly versified:

The feeble births acquired diseases chase,
Till Death extinguish the degenerate race. . . .
So, years successive, from perennial roots
The wire or bulb with lessen'd vigour shoots;
Till curlèd leaves, or barren flowers, betray
A waning lineage, verging to decay;
Or till, amended by connubial powers,
Rise seedling progenies from sexual flowers.
(II 165–6, 171–6)

Again he mentions the curious sexual habits of the Aphis:

Unmarried Aphides prolific prove
For nine successions uninform'd of love;
New sexes next with softer passions spring,
Breathe the fond vow, and woo with quivering wing.
(II 131–4)

Darwin introduces sexual reproduction, as the mechanism of the poem requires, by bringing on Cupid and Psyche. Hymen calls on everyone to praise them:

Behold, he cries, Earth! Ocean! Air above,
And hail the DEITIES OF SEXUAL LOVE!
All forms of Life shall this fond Pair delight,
And sex to sex the willing world unite. (II 243–6)

Darwin follows this up with many amusing examples of married bliss or jealousy.

The Lion-King forgets his savage pride,
And courts with playful paws his tawny bride;
The listening Tiger hears with kindling flame
The love-lorn night-call of his brinded dame.
Despotic LOVE dissolves the bestial war,
Bends their proud necks, and joins them to his car. . . .
Pleased as they pass along the breezy shore
In twinkling shoals the scaly realms adore,
Move on quick fin with undulating train,

Or lift their slimy foreheads from the main.
High o'er their heads on pinions broad display'd
The feather'd nations shed a floating shade;
Pair after pair enamour'd shoot along,
And trill in air the gay impassion'd song.

(II 357–62, 371–8)

They are followed, in lines 389–410, by Love's car, all flower-bedecked: it is the lightest of light verse, but too lengthy to quote here.

The third canto is devoted to the 'Progress of the Mind'. To start the action, Urania and the Muse make a round-the-world tour:

First with fond gaze blue fields of air they sweep,
Or pierce the briny chambers of the deep;
Earth's burning line, and icy poles explore,
Her fertile surface, and her caves of ore;
Or mark how Oxygen with Azote-Gas
Plays round the globe in one aerial mass,
Or fused with Hydrogen in ceaseless flow
Forms the wide waves, which foam and roll below.

(III 9–16)

After this quick inspection Urania tells the Muse (and the inevitable squadrons of 'attentive Nymphs') how Man's powers of reason developed. Darwin particularly emphasizes two physical features that helped in man's evolutionary development—hand and eye:

The branching forehead with diverging horns
Crests the bold bull, the jealous stag adorns. . . .
The tropic eel, electric in his ire,
Alarms the waves with unextinguish'd fire;
The fly of night illumes his airy way,
And seeks with lucid lamp his sleeping prey. . . .
Proud Man alone in wailing weakness born,
No horns protect him, and no plumes adorn;
No finer powers of nostril, ear, or eye,
Teach the young Reasoner to pursue or fly.—
Nerved with fine touch above the bestial throngs,
The hand, first gift of Heaven! to man belongs.

(III 103–4, 111–14, 117–22)

This fine touch allows an accuracy of outline, which the eye takes advantage of:

> Symbol of solid forms is colour'd light,
> And the mute language of the touch is sight. (III 143–4)

This leads on to Darwin's questionable theory of beauty—that a baby learns 'Ideal Beauty from its Mother's breast'. Darwin believes his theory explains our pleasure when we come across the serpentine line of beauty defined by Hogarth, or when

> the wide eye the wavy lawns explores,
> The bending woodlands, or the winding shores,
> Hills, whose green sides with soft protuberance rise,
> Or the blue concave of the vaulted skies, (III 207–10)

or when we see

> the nice curves which swell the female breast. (III 216)

All this, Darwin thinks, is attributable to early childhood memories.

Having exhausted this rather Freudian streak, Darwin offers a philosophy of art and science. They can be explained, he thinks, in terms of Imitation:

> Whence the fine power of IMITATION springs,
> And apes the outlines of external things;
> With ceaseless action to the world imparts
> All moral virtues, languages, and arts. (III 285–8)

His idea is that we merely imitate what we see, or what others have done; but occasionally, in the hands of a creative artist or scientist, the imitations lead on to a new synthesis.

> Hence to clear images of form belong
> The sculptor's statue, and the poet's song,
> The painter's landscape, and the builder's plan,
> And IMITATION marks the mind of Man. (III 331–4)

Darwin provides a pithy summary of Man's characteristics, not forgetting a dig at the then-widespread habit of praying for advantage:

> To express his wishes and his wants design'd
> Language, the *means*, distinguishes Mankind;

For *future* works in Art's ingenious schools
His hands unwearied form and finish tools;
He toils for money *future* bliss to share,
And shouts to Heaven his mercenary prayer. (III 437–42)

Though he disliked most religious practices, Darwin approved the emphasis on sympathy and benevolence in New Testament morality:

> The Seraph, SYMPATHY, from Heaven descends,
> And bright o'er earth his beamy forehead bends;
> On Man's cold heart celestial ardor flings,
> And showers affection from his sparkling wings. . . .
> Points with uplifted hand to realms above,
> And charms the world with universal love. . . .
> High on yon scroll, inscribed o'er Nature's shrine,
> Live in bright characters the words divine.
> 'IN LIFE'S DISASTROUS SCENES TO OTHERS DO,
> WHAT YOU WOULD WISH BY OTHERS DONE TO YOU.'
> (III 467–70, 477–8, 487–90)

This precept, 'if sincerely obeyed by all nations, would a thousand-fold multiply the present happiness of mankind', Darwin remarks in a note.

The fourth canto, 'Of Good and Evil', begins with the grim picture of the struggle for existence already quoted in Chapter 6. Then Darwin emphasizes the troubles of man, particularly the self-imposed ones, such as 'the curst spells of Superstition' which fetter 'the tortured mind'. Other forms of mental illness are run over with fine verve:

> There writhing Mania sits on Reason's throne,
> Or Melancholy marks it for her own,
> Sheds o'er the scene a voluntary gloom,
> Requests oblivion, and demands the tomb. . . .
> Here ragged Avarice guards with bolted door
> His useless treasures from the starving poor;
> Loads the lorn hours with misery and care,
> And lives a beggar to enrich his heir. . . .
> While mad with foolish fame, or drunk with power,
> Ambition slays his thousands in an hour.
> (IV 89–92, 97–100, 103–4)

Is there nothing but woe, then? On the contrary, Urania replies, good and evil are for man nicely balanced. On the side of good we have the joy of consciousness, the pleasures of Love, the delights of Nature, and the enjoyments of music, painting and all the imaginative arts. Then there are the satisfactions of philanthropy and the triumphs of science, from the sublime concepts of Newton, who

> Explored in Nature's scenes the effect and cause,
> And, charm'd, unravell'd all her latent laws,

to the practical worth of Archimedes, Savery's steam engine or Arkwright's mill. Darwin guesses, with remarkable foresight, that these triumphs of technology are but the prelude to skyscrapers, piped water and traffic jams:

> Bid raised in air the ponderous structure stand,
> Or pour obedient rivers through the land;
> With cars unnumber'd crowd the living streets,
> Or people oceans with triumphant fleets. (IV 315–18)

At the end Darwin returns to the struggle for existence and propounds his philosophy of organic happiness, in the lines already quoted in Chapter 6. Darwin puts his own philosophy into the mouth of Pythagoras:

> So erst the Sage with scientific truth
> In Grecian temples taught the attentive youth;
> With ceaseless change how restless atoms pass
> From life to life, a transmigrating mass;
> How the same organs, which to-day compose
> The poisonous henbane, or the fragrant rose,
> May with to-morrow's sun new forms compile,
> Frown in the Hero, in the Beauty smile.
> Whence drew the enlighten'd Sage the moral plan,
> That man should ever be the friend of man;
> Should eye with tenderness all living forms,
> His brother-emmets, and his sister-worms. (IV 417–28)

Darwin then offers a convincing picture of limestone deposits and coral islands being formed:

> The marble mountain, and the sparry steep,
> Were built by myriad nations of the deep,—

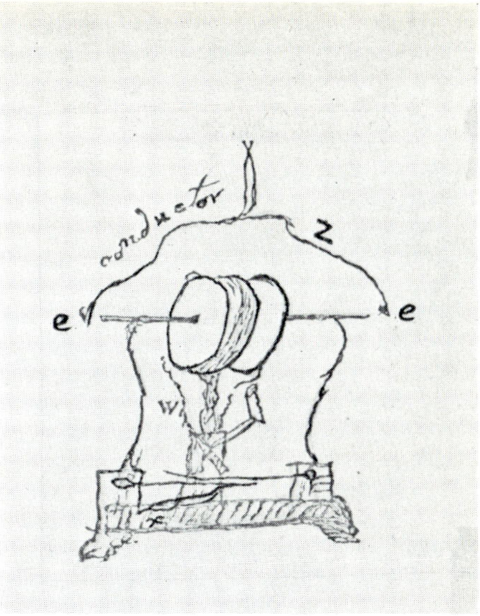

FIG. 13 (*right*). Darwin's electrical doubler, 1778, which was quite famous in its day. From a sketch in his Commonplace Book, slightly touched up. For further details, see page 127.

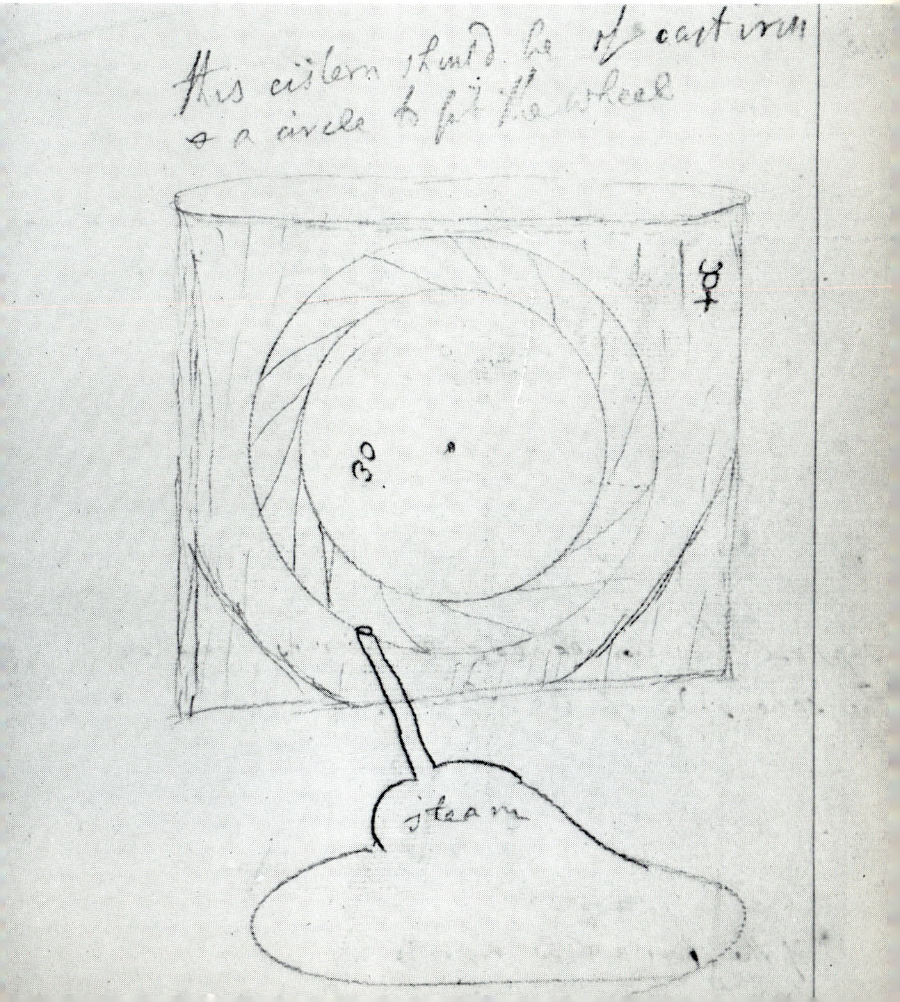

FIG. 14 (*below*). Sketch of steam turbine, from Darwin's Commonplace Book, 1779. The alchemist's symbol for water appears at the top right-hand corner of the cistern. Steam turbines of this type were not developed until late in the nineteenth century.

FIG. 15. The most elaborate of Darwin's many designs for raising water. From *Phytologia* (1800). For explanation, see page 128.

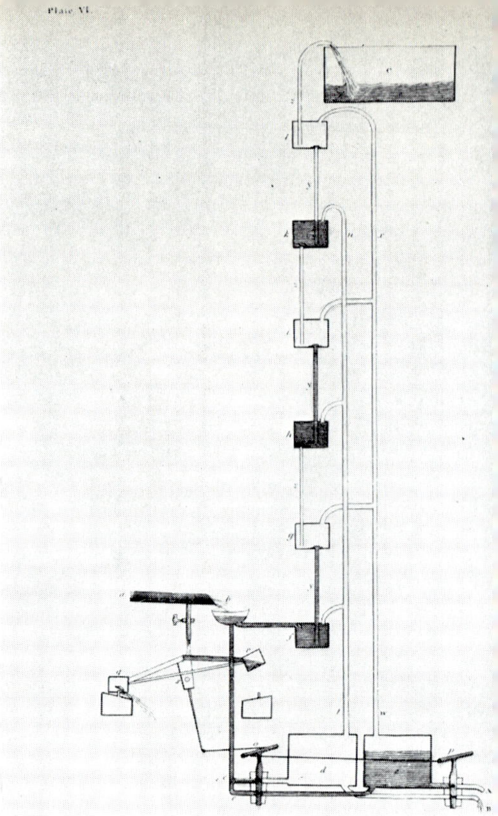

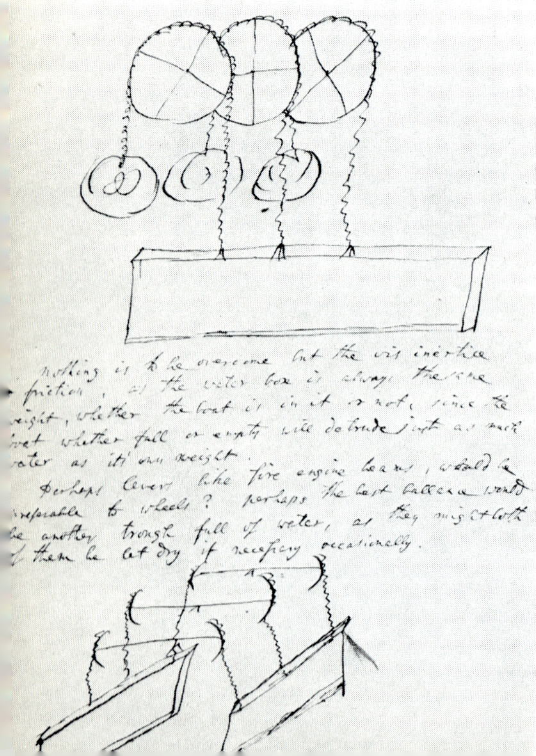

FIG. 16. Some of Darwin's sketches for a canal lift, from his Commonplace Book, 1777. For discussion, see page 129.

Age after age, who form'd their spiral shells,
Their sea-fan gardens and their coral cells;
Till central fires with unextinguished sway
Raised the primeval islands into day;—
The sand-fill'd strata stretch'd from pole to pole;
Unmeasured beds of clay, and marl, and coal,
Black ore of manganese, the zinky stone,
And dusky steel on his magnetic throne,
In deep morass, or eminence superb,
Rose from the wrecks of animal or herb. (IV 431–42)

This is followed by the picture of the mountains as mighty monuments of past delight (see Chapter 6).

The poem concludes with due ceremony, as the Nymphs gather for a graceful choral finale:

Fair virgin trains in bright procession move,
Trail their long robes, and whiten all the grove;
Pair after pair to Nature's temple sweep,
Thread the broad arch, ascend the winding steep;
Through brazen gates along susurrant ailes
Stream round their GODDESS the successive files.
(IV 473–8)

Finally Urania scatters incense on Nature's altar and

Lifts her ecstatic eyes to TRUTH DIVINE!

Darwin's reputation as a poet declined steadily between 1800 and 1850, but even the remnants of that reputation were impressive: for example, G. L. Craik, in his popular *History of English Literature* in the 1860s, was hostile to Darwin but still gave him twice as much space as Shakespeare. Between 1870 and 1950 Darwin's literary achievements were neglected.

The revival of his reputation began in the 1950s. If we want a precise date, 1954 is as good as any, for in that year a modern annotated edition of *The Temple of Nature* was published—the first since 1825—and it proved so popular that a second edition followed in 1960. It would be unfair to conceal the fact that these editions were of a Russian translation, by N. A. Kholodkovskii, with extensive and well-informed notes by E. N. Pavlovsky. But before long the prophet was being honoured even in his own land, or at least in his own language. In the past ten years the importance of his poems has been fully recognized, and his work has attracted the attention of a number of literary critics and historians of ideas—Elizabeth Sewell, Bernard Blackstone, H. W. Piper, Irwin Primer and others.[36]

11 Romantic Followers

Dr. Darwin's Poetry . . . arrests the attention too often.
COLERIDGE; *Notebook*

DARWIN'S poems were immensely popular in the early 1790s, and showed that Nature, science and radical ideas could all be brought into a poem. *The Loves of the Plants* was published in the crucial revolutionary year, 1789:

Bliss was it in that dawn to be alive,
But to be young was very heaven,

wrote Wordsworth, who was nineteen at the time. The French Revolution had a powerful effect on Wordsworth; yet for him, and for Coleridge too, *The Botanic Garden* had a longer and a stronger influence. Darwin's sentient plants are at the very root of Wordsworth's credo,

'Tis my faith that every flower
Enjoys the air it breathes.

During the 1790s both Wordsworth and Coleridge came utterly to reject Darwin's polished couplets, and the preface to *Lyrical Ballads* is one long blast against Darwin. But they accepted much of his subject-matter and ideas, even though they despised his style. Shelley and Keats were also greatly indebted to Darwin, and many critics now seem to accept that Darwin was the most important single influence on the English Romantic poets. It is not appropriate to argue this thesis here; it is discussed in the books already cited.[36] All I shall do is to quote some of the passages in Darwin's poems which clearly influenced four of the leading Romantic poets, Wordsworth, Coleridge, Keats and Shelley.

As a preliminary, I must comment on *The Loves of the Triangles*, published in 1798. In the hectic 'anti-Jacobin' fever then prevailing, the political and religious radicalism of Darwin and of William Godwin were obnoxious to the Government; the popularity of Darwin's poems and Godwin's *Political Justice* was alarming. Canning, who was then

Foreign Under-Secretary in Pitt's government, had the idea of destroying Darwin's reputation with a parody and ridiculing Godwin at the same time. With two collaborators, Hookham Frere and George Ellis, he wrote *The Loves of the Triangles*, published in the *Anti-Jacobin*.

At its best *The Loves of the Triangles* is just as amusing as *The Botanic Garden*, sometimes even more so, as when the Nymph of the Wheel eyes her beloved Smoke-Jack:

> Mark how his various parts together tend,
> Point to one purpose,—in one object end:
> The spiral *grooves* in smooth meanders flow,
> Drag the long *chain*, the polish'd axles glow,
> While slowly circumvolves the piece of beef below:
> The conscious fire with bickering radiance burns,
> Eyes the rich joint, and roasts it as it turns.
> (*Poetry of the Anti-Jacobin*, Blackwell, 1924, p. 95)

But the basis of the parody is rather childish, and the antics of the amorous triangles and rectangles are more tedious than those of Darwin's youths and virgins. The real sting of the *Loves of the Triangles* lay in its sarcastic notes: these ridiculed Darwin's ideas that (i) man's remote ancestors were lower animals, (ii) electricity might have important uses, and (iii) the Earth is older than the Bible says. Since all three of Darwin's ideas were right, the parodists fall flat on their facetious faces. At the time, however, no one knew they were wrong, and so they were able to undermine Darwin's reputation, greatly blunting the impact of *The Temple of Nature* in 1803.

Darwin's influence lived on, however, as we shall see from the following parallels between his works and that of the Romantic poets. Every poet stores away in his memory images and phrases he has read, and they slip out unconsciously into his own poems. It is no disgrace to the poet that we can find parallels with earlier work, in this instance Darwin's; but such parallels do suggest that the earlier work won its way into his unconscious and thereby influenced him quite deeply.

Darwin proved useful to Wordsworth in several ways. In his first long poem, *Descriptive Sketches* (1793), Wordsworth adopted Darwin's style, and then, a few years later, rebelled against it, in the revolutionary *Lyrical Ballads*. Secondly, Darwin was an ardent champion of the French Revolution, which Wordsworth thought so glorious. Thirdly, Wordsworth came to adopt a religious creed very similar to Darwin's, including the idea of a vague Power lying behind the processes of Nature, and as much concerned with Nature as with Man. The final resemblance is that most of Wordsworth's poems are concerned with Nature, strong on exact description, and particularly concentrated on flowers: this is just the recipe of *The Botanic Garden*.

There are a few specific resemblances to Darwin in Wordsworth's poems. The end of Darwin's account of Cambyses's army being buried in sand,

awhile the living hill
Heaved with convulsive throes—and all was still!

is reproduced in Wordsworth's *To Enterprise* thirty years later:

An Army now, and now a living hill
That a brief while heaves with convulsive throes—
Then all is still.
(*Economy of Vegetation* II 497–8; *To Enterprise*, 114–16)

A second example is Darwin's story of the farmer cursed by an old woman stealing sticks, who told him he would never be warm again (*Zoonomia*, IV 68–9). Wordsworth versifies this story in *Goody Blake and Harry Gill*, which begins with the much-parodied lines about the chilly Harry:

Oh! what's the matter? what's the matter?
What is 't that ails young Harry Gill?
That evermore his teeth they chatter,
Chatter, chatter, chatter still!
Of waistcoats Harry has no lack,
Good duffle grey, and flannel fine;
He has a blanket on his back,
And coats enough to smother nine.

Other possible parallels are Darwin's

Hang o'er the gliding steel, and hiss along the ice,

and Wordsworth's

All shod with steel,
We hissed along the polished ice.
(*Economy of Vegetation*, III 570; *Prelude* I 433–4)

And, much more questionably, Darwin's

All wan and shivering in the leafless glade

and Wordsworth's famous line,

All bright and glittering in the smokeless air.[37]
(*Loves of the Plants* I 317; *Westminster Bridge*, 8)

If Wordsworth offers few verbal parallels, Coleridge provides far too many. He admitted that Darwin's verse arrested his attention too often, and 'once on the trail of Coleridge through Erasmus Darwin, and especially *The Botanic Garden*, one can find almost anything one is looking for', as Coleridge's editor, Kathleen Coburn, remarks.[38]

Take first the most famous of Coleridge's poems, the *Ancient Mariner*, written in 1797–8, about a year after that long talk with Darwin which he found so rewarding—on every subject but religion. The climax of Coleridge's poem comes when the Mariner looks at the luminous creatures in the wake of the ship and suddenly feels a universal sympathy with all living creatures. Immediately the Albatross falls from his neck and he sheds his guilt. The Mariner's sympathy with the whole of organic nature—

He prayeth well, who loveth well
Both man and bird and beast—

is precisely Darwin's philosophy of organic happiness, which requires that man 'should eye with tenderness all living forms'. Darwin is also probably responsible for the particular scene which affected the Mariner, for he mentions the luminous trails of ships in *The Economy of Vegetation*,

Or gild the surge with insect-sparks, that swarm
Round the bright oar, the kindling prow alarm,

to which he adds the note:

In some seas, as particularly about the coast of Malabar, as a ship floats along, it seems during the night to be surrounded with fire, and to leave a long tract of light behind it. Whenever the sea is gently agitated it seems converted into little stars, every drop as it breaks emits light, like bodies electrified in the dark.

Coleridge describes the snaky tracts of light as follows:

Beyond the shadow of the ship,
I watched the water-snakes:
They moved in tracks of shining white,
And when they reared, the elfish light
Fell off in hoary flakes.
(*Economy of Vegetation* I 199–200
and Note IX; *Ancient Mariner*, 272–6)

Darwin's powerful lines about the polar regions—

Slow o'er the printed snows with silent walk
Huge shaggy forms across the twilight stalk;
And ever and anon with hideous sound
Burst the thick ribs of ice, and thunder round—

seem to be re-echoing in the *Ancient Mariner*:

Nor shapes of men nor beasts we ken—
The ice was all between.

The ice was here, the ice was there,
The ice was all around:
It cracked and growled, and roared and howled,
Like noises in a swound. . . .
The ice did split with a thunder-fit.
(*Loves of the Plants*, IV 349–52; *Ancient Mariner*, 57–62, 69)

Darwin has several descriptions of the aurora, and comments on the stars shining through aurorae and the tails of comets, his line

The wan stars glimmering through its silver train

being accompanied by a note that

small stars are seen undiminished through both the light of the tails of comets, and of the aurora borealis.

Coleridge also brings in the aurora, and the 'wan stars' shining through:

> The upper air burst into life. . . .
> And to and fro, and in and out,
> The wan stars danced between.
> (*Economy of Vegetation*, I 134 and Note IV;
> *Ancient Mariner* 313, 316–7)

Kubla Khan also has its echoes of Darwin. His description of the river Liffey,

> So from secluded springs, and secret caves,
> Her Liffy pours his bright meandering waves. . . .
> Spread the bright treasure in a thousand rills. . . .
> And blossom'd orchards stretch their silver shade,

has its parallel in the sacred river Alph, running

> Through caverns measureless to man. . . .
> . . . meandering with a mazy motion. . . .
> [through] gardens bright with sinuous rills,
> Where blossomed many an incense-bearing tree.
> (*Economy of Vegetation*, III 459–60, 520, 468;
> *Kubla Khan*, 4, 25, 8–9)

And Darwin's account of the 'glittering domes' in Cracow's mighty caverns, from which

> Impetuous fountains burst their headlong way,

recalls 'that deep romantic chasm' (not far from the pleasure-dome and the caverns), from which

> A mighty fountain momently was forced.
> (*Economy of Vegetation*, II 130; *Kubla Khan*, 19)

Another of Coleridge's poems with many Darwinian echoes is *The Destiny of Nations* (1796). Here are two:

> Or yoke the vollied lightnings to your cars (Darwin);
> Yoke the red lightnings to their volleying car (Coleridge);

and

> When LOVE DIVINE, with brooding wings unfurl'd,
> Call'd from the rude abyss the living world (Darwin);

When Love rose glittering, and his gorgeous wings
Over the abyss fluttered with such glad noise (Coleridge).
(*Economy of Vegetation*, I 116 and 101–2; *Destiny of Nations*, 54 and 283–4. Cf. also *Paradise Lost*, VII 233–5)

The Economy of Vegetation may have nauseated Coleridge; but after all these examples we can scarcely doubt that he read it attentively.

Even an ill wind was welcomed when it came from *The Economy of Vegetation*: Darwin tells how the stifling desert wind, the Simoom,

rides the tainted air,
Points his keen eye, and waves his whistling hair.

Coleridge follows in *Religious Musings* with

through the tainted noon
The Simoom sails.
(*Economy of Vegetation* IV 67–8; *Religious Musings*, 268–9)

After so many unconscious resemblances, it is quite a relief to find a conscious (though still unacknowledged) borrowing: a 25-line footnote to Coleridge's *Lines at Shurton Bars*, on the subject of 'Light from plants', is copied word for word from Darwin's account in a note to *The Loves of the Plants* (pp. 182–3). The only change Coleridge makes is to introduce a few misprints.[39]

Of the leading Romantic poets, Keats might seem to have had most in common with Darwin: Keats had training as a doctor, and he revelled in the sensuous and in 'the realm of Flora and old Pan'. If we can rely on Mrs SchimmelPenninck's account of the meals laid on for Darwin, the feast in *The Eve of St. Agnes* is just what the doctor ordered:

candied apple, quince, and plum, and gourd;
With jellies soother than the creamy curd,
And lucent syrops, tinct with cinnamon;
Manna and dates, in argosy transferr'd
From Fez. . . .

Bernard Blackstone in his book *The Consecrated Urn* has suggested that the entire poetic career of Keats parallels

Darwin's: 'The journey we go from the calixes of "small buds unfolding" in the early *Sleep and Poetry* to the ultimate *Grecian Urn* is the same journey we go from the opening plates of urn-like seed-cases and flowers in *The Loves of the Plants* to the concluding engravings of the Portland Vase in *The Economy of Vegetation*'. (p. 337)

Keats, however, like Wordsworth, only borrows a few specific images from Darwin. Prominent among these are Darwin's trains of virgins moving towards flowered altars, as, for example, in *The Temple of Nature*:

> Long trains of virgins from the sacred grove,
> Pair after pair, in bright procession move,
> With flower-fill'd baskets round the altar throng,
> Or swing their censers, as they wind along.

Similar censer-swinging virgins appear in Keats's *Ode to Psyche*:

> Nor altar heap'd with flowers;
> Nor virgin-choir to make delicious moan. . . .
> From swingèd censer teeming;
> Thy shrine, thy grove, thy oracle. . . .
> (*Temple of Nature*, I 159–62; *Ode to Psyche*, 29–30, 47–8)

Keats also seems to have been impressed by Darwin's chilling picture of a church by night in *The Loves of the Plants*:

> Hoarse on their hinge the ponderous portals jar;
> As through the colour'd glass the moon-beam falls,
> Huge shapeless spectres quiver on the walls;
> Low murmurs creep along the hollow ground,
> And to each step the pealing ailes resound;
> By glimmering lamps, protecting saints among,
> The shrines all trembling as they pass along.

Keats seems to follow in his footsteps in *The Eve of St. Agnes*, first when the Beadsman returns

> Along the chapel aisle by slow degrees:
> The sculptur'd dead, on each side, seem to freeze;

and again, later in the poem, when Porphyro stands

beside the portal doors,
Buttress'd from moonlight,

and when the moon-beams fall through the casement

diamonded with panes of quaint device,
Innumerable of stains and splendid dyes,

to shine with warm colours on the sleeping Madeline.*

Another line of Darwin's,

By each cold nymph her marble lover lies,

possibly contributed both to the Beadsman's idea that the 'sculptured dead . . . may ache in icy hoods and mails', and to the image of the 'brede of marble men and maidens' in the *Ode on a Grecian Urn*, a poem which might be a response to Darwin's plea,

bid Mortality rejoice and mourn
O'er the fine forms on PORTLAND's mystic urn.
(*Loves of the Plants*, II 289; *Economy of Vegetation*, II 319–20)

Many single lines in Darwin's poems seem 'Keatsian'. For example Canto I, line 76, of the *Economy of Vegetation* reads,

On winnowing wings and waft her golden hair.

This is not far removed from

Thy hair soft-lifted by the winnowing wind,

in the ode *To Autumn*. Three lines later Darwin comes up with

Musk'd in the rose's lap fresh dews they shed,

which is redolent of

The coming musk-rose full of dewy wine,

in the *Ode to a Nightingale*. These resemblances show how Keats breathed the atmosphere of Darwin's poems, even though he rejected their stiff couplets.[40]

* *Loves of the Plants*, III 22–8, *Eve of St. Agnes*, 13–14, 76–7, 211–12.

The last of the four poets, Shelley, was by far the keenest follower of Darwin. From Darwin he had the idea of combining science and poetry, and in poems like *The Cloud* Shelley eventually achieved a fusion of poetry and science which went far beyond Darwin's more mechanical skills. Shelley, like Darwin, was a sceptic in religion, a keen radical in politics, and a fighter against tyranny and oppression in all its forms. Darwin's philosophy of organic happiness also appealed to Shelley, partly because he liked the idea of plants having 'feelings'.

Shelley's early poems, first published in 1964, have many echoes of Darwin.[41] Perhaps the most striking, because it is so ludicrous, derives from Darwin's story of a determined dame called Eliza seeking her husband at the battle of Minden:

> So stood Eliza on the wood-crown'd height,
> O'er Minden's plain, spectatress of the fight,
> Sought with bold eye amid the bloody strife
> Her dearer self, the partner of her life. . . .
> Near and more near the intrepid Beauty press'd. . . .
> O'er groaning heaps, the dying and the dead. . . .
> So wings the wounded Deer her headlong flight. . . .

The Shelleyan parallel comes in the most substantial of his early poems, *Henry and Louisa*:

> When frantic o'er the waste Louisa sped
> To drink her dying lover's latest vow.
> Sighed 'mid her locks the sea-gales as they blew,
> Bearing along faint shrieks of dying men. . . .
> Silent she paused a space, and then again
> New-nerved by fear and hope, sprang wild across the plain. . . .
> Fleet as the wild deer, his Louisa came.*

Shelley's first long poem, *Queen Mab*, is a tirade against present evils, and its young author closely follows Darwin's attacks in *The Temple of Nature* on superstition, tyrants, slavery, war, alcohol, avarice and luxury, as well as using his

* *Loves of the Plants*, III 269–72, 281, 308, 263; *Henry and Louisa*, 214–17, 219–20, 233 (*The Esdaile Poems*, ed. N. Rogers, O.U.P., 1966, pp. 88–9).

machinery of a Goddess in a 'magic car'. Here are four of the neatest verbal parallels:

While mad with foolish fame, or drunk with power,
Ambition slays his thousands in an hour. (Darwin)
When merciless ambition, or mad zeal,
Has led two hosts of dupes to battlefield. (Shelley)

And one great Slaughter-house the warring world! (Darwin)
Making the earth a slaughter-house! (Shelley)

when a Monarch or a mushroom dies. (Darwin)
Yon monarch, in his solitary pomp,
Was but the mushroom of a summer day. (Shelley)

The Lion-King forgets his savage pride,
And courts with playful paws his tawny bride. (Darwin)
The lion now forgets to thirst for blood:
There might you see him sporting in the sun. (Shelley)
(*Temple of Nature*, IV 103–4; IV 66; IV 383; II 357–8.
Queen Mab, VI 178–9; VII 48: IX 31–2; VIII 124–5)

As a second example we may take Shelley's greatest poem, *Prometheus Unbound*, whose main theme is the regeneration of the world through Universal Love. Darwin has a rather similar picture in the lines already quoted from Canto III of *The Temple of Nature*, when the seraph Sympathy 'charms the world with universal love', and again in *The Economy of Vegetation* (IV 62), when 'Love and Beauty rule the willing world'.

There are also plenty of specific verbal echoes in *Prometheus Unbound*. For example Darwin's rather stilted line,

So turns the needle to the pole it loves,

is altered by Shelley into an attractive couplet:

Like the polar Paradise,
Magnet-like of lovers' eyes.
(*Loves of the Plants*, I 315; *Prometheus Unbound*, IV 465–6)

Darwin's dance of the Nereids,

In air and ocean weave the mazy dance,

foreshadows Shelley's

Weave the dance on the floor of the breeze.
(*Loves of the Plants*, IV 218; *Prometheus Unbound*, IV 69)

And Darwin's

Hung with gold-tresses o'er the vast inane

provides the cue for the climactic line which ends Act III of *Prometheus Unbound*:

Pinnacled dim in the intense inane.
(*Economy of Vegetation*, I 98; *Prometheus Unbound*, III 204)

Shelley's most popular poems are his lyrics of the sky, the *West Wind*, the *Skylark* and *The Cloud*. Darwin was keen on aerial phenomena and provides some parallels, as when he asks the Nymphs to unbind 'the WESTERN WIND'

And wring the rain-drops from his tangled hair.

Shelley also personifies the West Wind, who

Shook from the tangled boughs of Heaven and Ocean,
Angels of rain and lightning.
(*Economy of Vegetation*, I 432; *Ode to the West Wind*, 17–18)

Later in the *Economy of Vegetation* Darwin shows us Jove as he retires with Juno on the couch imperial,

And rests the crimson cushions upon clouds.
Earth feels the grateful influence from above,
Sighs the soft Air, and Ocean murmurs love;
Etherial Warmth expands his brooding wing,
And in still showers descends the genial Spring.

This is quite like Shelley's Cloud:

And when Sunset may breathe, from the lit sea beneath,
  Its ardours of rest and of love,
And the crimson pall of eve may fall
  From the depth of Heaven above,
With wings folded I rest, on mine aery nest,
  As still as a brooding dove.
(*Economy of Vegetation*, III 256–60; *The Cloud*, 39–44)

And Darwin's phrase,

each nice pore of ocean, earth and air,

inspired Shelley's brilliant summary of water circulation,

> I pass through the pores of the ocean and shores.
> (*Economy of Vegetation*, I 85; *The Cloud*, 75)

Shelley's *Sensitive Plant*, with its catalogue of continuously personified flowers, plainly derives from the *Loves of the Plants*. Shelley's *Adonais* has striking affinities with Canto II of *The Temple of Nature*: for in both Urania laments over human wrongs. Shelley's *Epipsychidion* has several images taken from Darwin. About a dozen further such parallels have been noted.[42]

Although I have only given examples from four of the leading poets, many other writers of the time were influenced by Darwin. They include Blake, Byron, Crabbe, Landor and Mary Shelley.[43]

When a later poet unconsciously reproduces Darwin's phrases, we can be fairly sure he was influenced by Darwin. But there is room for argument on how exact the reproduction must be, and a few of the examples I have quoted may be merely due to chance. Since Darwin's verse was so effective in 'arresting the attention', as Coleridge complained, the resemblances seem more likely to be real than random. However, as Darwin himself said, 'the candid reader must determine'.

12 Research Worker

> Then must thou needs find out new heaven, new earth.
>
> SHAKESPEARE; *Antony and Cleopatra*

WE HAVE already seen something of Darwin's mechanical inventions in Chapter 8: now we look at what would today be called his scientific research.

Darwin wrote seven formal scientific papers, six of them published in the *Philosophical Transactions* of the Royal Society and the seventh in the *Medical Transactions* of the Royal College of Physicians.* In addition several of the notes to his poems are in effect scientific papers, and deserve to be discussed.

The first of his papers for the *Philosophical Transactions*, published in 1757, was entitled 'Remarks on the Opinion of Henry Eeles, Esq., concerning the Ascent of Vapour'. Eeles contended that vapours rise only if they are electrically charged. Darwin rightly refutes him, and maintains that the motions of vapours do not depend on electricity. However it so happened that in the late 1750s the fallacious electrical theories of rain propounded by the Italian scientist Beccaria were becoming popular, and Darwin's denial went almost unheeded.[44]

The second paper, 'An uncommon Case of an Haemoptysis' (1760), is not of much interest today; but these first two papers were enough to secure Darwin's election as a Fellow of the Royal Society in 1761, the beginning of a unique succession of five generations of Darwins who became Fellows of the Society.

The third paper, 'Experiments on Animal Fluids in the exhausted Receiver' (1774), is the one sent with the letter to Franklin quoted in Chapter 2. Darwin describes how blood swells up in a vacuum and suggests that deafness may sometimes be cured by putting a cupping glass over the ear and exhausting the air.

* For the titles of the papers, see the Book List, p. 205.

The fourth paper, 'A New Case in Squinting', takes us from deaf ears to blind spots. The subject is a five-year-old boy, Master Sandford, who looked at every object with only one eye (the one furthest from the object) by turning his head so that the image fell on the blind spot in the retina of the nearer eye. Darwin believed the habit was caused by his wearing, as a baby, a nightcap which had appendages at each side visible only to the opposite eye. Darwin's proposed cure was to fit a large false nose like the gnomon of a sundial, which would force the boy to view with the near eye. This treatment was not persevered with—presumably it was too unsightly—and when Darwin saw the boy again six years later he found the habit more fixed. He made experiments which showed that Master Sandford had blind spots four times larger than normal:

> Two circular papers, each of 4 inches diameter, were stuck against the wall, their centres being exactly at 8 inches distance from each other. On closing one eye, and viewing the central spot of these papers with the eye farthest from it, and then retreating 26 inches from it, the other paper became invisible. This experiment was made on 5 people, of various ages, from 10 years old to 40; and the paper disappeared to them all at about this distance, or an inch or 2 more or less; but to Master Sandford the paper disappeared at about 13 inches distance from the wall. (*Philosophical Transactions*, 68, 94–5 (1778))

This experiment is quite simple, but it does show how Darwin would take a great deal of trouble to elucidate a puzzling case, even though in this instance he was confident that the use of the gnomon would eventually effect a cure.

The fifth paper, 'An Account of an artificial Spring of Water' (1785), is an important one, since it contains the first clear exposition of the principle of the artesian well and also reveals Darwin's energy and skill as a constructional engineer. I quote the paper in full, except for the final paragraph.

> To the President and Fellows of the Royal Society.
>
> Gentlemen, Derby, July 16, 1784.
>
> Confident that every atom which may contribute to increase the treasury of useful knowledge, which you are so successfully

endeavouring to accumulate, will be agreeable and interesting to the Society, I send you an account of an artificial spring of water, which I produced last summer near the side of the river Darwent in Derby.

Near my house was an old well, about one hundred yards from the river, and about four yards deep, which had been many years disused on account of the badness of the water, which I found to contain much vitriolic acid, with, at the same time, a slight sulphureous smell and taste; but did not carefully analyse it. The mouth of this well was about four feet above the surface of the river; and the ground, through which it was sunk, consisted of a black, loose, moist earth, which appeared to have been very lately a morass, and is now covered with houses built upon piles. At the bottom was found a bed of red marl, and the spring, which was so strong as to give up many hogsheads in a day, oozed from between the morass and the marl: it lay about eight feet beneath the surface of the river, and the water rose within two feet of the top of the well.

Having observed that a very copious spring, called Saint Alkmund's well, rose out of the ground about half a mile higher on the same side of the Darwent, the level of which I knew by the height of the intervening wier to be about four or five feet above the ground about my well; and having observed that the higher lands, at the distance of a mile or two behind these wells, consisted of red marl like that in the well; I concluded that, if I should bore through this stratum of marl, I might probably gain a water similar to that of St. Alkmund's well, and hoped that at the same time it might rise above the surface of my old well to the level of St. Alkmund's.

With this intent a pump was first put down for the purpose of more easily keeping dry the bottom of the old well, and a hole about two and a half inches diameter was then bored about thirteen yards below the bottom of the well, till some sand was brought up by the auger. A wooden pipe, which was previously cut in a conical form at one end, and armed with an iron ring at the other, was driven into the top of this hole, and stood up about two yards from the bottom of the well, and being surrounded with well-rammed clay, the new water ascended in a small stream through the wooden pipe.

Our next operation was to build a wall of clay against the morassy sides of the well, with a wall of well-bricks internally, up to the top of it. This completely stopped out every drop of the old water; and, on taking out the plug which had been put in the wooden pipe, the new water in two or three days rose up to the top, and flowed over the edges of the well.

Afterwards, to gratify my curiosity in seeing how high the new spring would rise, and for the agreeable purpose of procuring the water at all times quite cold and fresh, I directed a pipe of lead, about eight yards long, and three-quarters of an inch diameter, to be introduced through the wooden pipe described above, into the stratum of marl at the bottom of the well, so as to stand about three feet above the surface of the ground. Near the bottom of this leaden pipe was sewed, between two leaden rings or flanches, an inverted cone of stiff leather, into which some wool was stuffed to stretch it out, so that, after having passed through the wooden pipe, it might completely fill up the perforation of the clay. Another leaden ring or flanch was soldered round the leaden pipe, about two yards below the surface of the ground, which, with some doubles of flannel placed under it, was nailed on the top of the wooden pipe, by which means the water was perfectly precluded from rising between the wooden and the leaden pipes.

This being accomplished, the bottom of the well remained quite dry, and the new water quickly rose about a foot above the top of the well in the leaden pipe; and, on bending the mouth of this pipe to the level of the surface of the ground, about two hogsheads of water flowed from it in twenty-four hours, which had similar properties with the water of St. Alkmund's well, as on comparison both these waters curdled a solution of soap in spirit of wine, and abounded with calcareous earth, which was copiously precipitated by a solution of fixed alkali, but the new water was found to possess a greater abundance of it, together with numerous small bubbles of aerial acid or calcareous gas.

The new water has now flowed about twelve months, and, as far as I can judge, is already increased to almost double the quantity in a given time; and from the rude experiments I made, I think it is now less replete with calcareous earth, approaching gradually to an exact correspondence with St. Alkmund's well, as it probably has its origin between the same strata of earth.

As many mountains bear incontestible marks of their having

been forcibly raised up by some power beneath them; and other mountains, and even islands, have been lifted up by subterraneous fires in our own times, we may safely reason on the same supposition in respect to all other great elevations of ground. Proofs of these circumstances are to be seen on both sides of this part of the country; whoever will inspect, with the eye of a philosopher, the lime-mountain at Breedon, on the edge of Leicestershire, will not hesitate a moment in pronouncing that it has been forcibly elevated by some power beneath it; for it is of a conical form, with the apex cut off, and the strata which compose the central parts of it, and which are found nearly horizontal in the plain, are raised almost perpendicularly, and placed upon their edges, while those on each side decline like the surface of the hill; so that this mountain may well be represented by a bur made by forcing a bodkin through several parallel sheets of paper. At Router, or Eaglestone, in the Peak, several large masses of grit-stone are seen on the sides and bottom of the mountain, which by their form evince from what parts of the summit they were broken off at the time it was elevated; and the numerous loose stones scattered about the plains in its vicinity, and half buried in the earth, must have been thrown out by explosions, and prove the volcanic origin of the mountain. Add to this the vast beds of toad-stone or lava in many parts of this county, so accurately described, and so well explained, by Mr. Whitehurst, in his Theory of the Formation of the Earth.

Now as all great elevations of ground have been thus raised by subterraneous fires, and in a long course of time their summits have been worn away, it happens that some of the more interior strata of the earth are exposed naked on the tops of mountains; and that, in general, those strata which lie uppermost, or nearest to the summit of the mountain, are the lowest in the contiguous plains. This will be readily conceived if the bur, made by thrusting a bodkin through several parallel sheets of paper, had a part of its apex cut off by a penknife, and is so well explained by Mr. Michell, in an ingenious paper on the Phaenomena of Earthquakes, published a few years ago in the Philosophical Transactions.

And as the more elevated parts of a country are so much colder than the vallies, owing, perhaps, to a concurrence of two or three causes, but particularly to the less condensed state of the air upon hills, which thence becomes a better conductor of heat, as well as of electricity, and permits it to escape the faster; it is from the water condensed on these cold surfaces of mountains that our common cold springs have their origin; and which, sliding between two of the strata above described, descend till they find or make themselves an outlet, and will in consequence rise to a level with the part of the mountain where they originated. And hence, if by piercing the earth you gain a spring between the second and third, or third and fourth stratum, it must generally happen that the water from the lowest stratum will rise the highest, if confined in pipes, because it comes originally from a higher part of the country in its vicinity.

The increasing quantity of this new spring, and its increasing purity, I suppose to be owing to its continually dissolving a part of the earth it passes through, and hence making itself a wider channel, and that through materials of less solubility. Hence it is probable that the older and stronger springs are generally the purer; and that all springs were originally loaded with the soluble impurities of the strata through which they transuded.

[The final paragraph is a plea for the wider use of artesian wells for supplying water to houses and for irrigation. It is not quoted because it is very similar to the quotation from *Phytologia*, on page 112] (*Philosophical Transactions*, 75, 1–6 (1785).)

Chronologically the next of Darwin's papers is that in the *Medical Transactions*, 'An Account of the successful Use of Foxglove in some Dropsies'. I shall not discuss this paper, however, because it raises the controversial issue of whether Darwin or Withering deserves credit for first using digitalis to treat dropsy. My own view is that neither of them deserves much credit, since the treatment was known to Galen 1,600 years before.[45]

Darwin's last paper in the *Philosophical Transactions*, in 1788, contains his most important contributions to physical science. The full title of the paper is 'Frigorific Experiments on the mechanical Expansion of Air, explaining the Cause of the great Degree of Cold on the Summits of high Mountains, the sudden Condensation of aerial Vapour, and of the perpetual Mutability of atmospheric Heat'.

Two separate advances are marked by this paper, the first in physics, the second in meteorology. First, Darwin gives a clear description of the process of adiabatic expansion, the way in which air cools if it is allowed to expand in the absence of external sources of heat or cold. This process is fundamental in all heat engines and in the atmosphere: instances had been noted previously, but Darwin seems to have been the first to expound adiabatic expansion as a principle. However, he failed to treat the subject numerically: the mathematical relationship between temperature T and pressure p in adiabatic expansion*, which is as fundamental as Boyle's Law, was not fully understood until after Joule's work in the 1840s.[46]

Darwin's second and greater achievement was to explain how most clouds form, by applying the adiabatic principle: 'when large districts of air . . . are raised . . . they become . . . so much expanded by the great diminution of the pressure over them . . . that . . . the air . . . by its expansion produces cold and devaporates'. *Devaporate* is a word coined by Darwin, meaning *condense into droplets*: it is a great pity that meteorologists have failed to adopt Darwin's word, since (as Knowles Middleton has pointed out) it is much better than the normal *condense*, which is very ambiguous.[47]

Since this is the most important of Darwin's scientific papers, I quote it in full, apart from two paragraphs:

> Having often revolved in my mind the great degree of cold producible by the well known experiments on evaporation; in which, by the expansion of a few drops of ether into vapour, a thermometer may be sunk much below the freezing point; and recollecting at the same time the great quantity of heat which is necessary to evaporate or convert into steam a few ounces of boiling water; I was led to suspect that elastic fluids, when they were mechanically expanded, would attract or absorb heat from the bodies in their vicinity; and that, when they were mechanically condensed, the fluid matter of heat would be pressed out of them, and diffused among the adjacent bodies.
>
> As this principle might possibly be extended to elastic solid bodies, as well as to fluid ones, and explain the cause of the heat occasioned by percussion or friction, and by some chemical

* The relationship is $p/p_0 = (T/T_0)^{\gamma/(\gamma-1)}$, where γ is the ratio of specific heats and suffix o denotes initial values.

combinations, as well as the perpetual mutability of it in the atmosphere, I have, at different times, endeavoured to subject it to experiment.

1. When Dr. Hutton of Edinburgh, and Mr. Edgeworth of Edgeworthtown in Ireland, were with me about twelve or fourteen years ago, the following experiment, which had been proposed by one of the company, was carefully made. The blast from an air-gun was repeatedly thrown on the bulb of a thermometer, and it uniformly sunk it about two degrees. The thermometer was firmly fixed against a wall, and the air-gun, after being charged, was left for an hour in its vicinity, that it might previously lose the heat acquired in the act of charging; the air was then discharged in a continued stream on the bulb of the thermometer, and the event shewed that the air at the time of its expansion attracted or absorbed heat from the mercury of the thermometer.

In March 1785, by the assistance of Mr. Fox and Mr. Strutt, of Derby, a thermometer was fixed in a wooden tube, and so applied to the receiver of an air-gun, that, on discharging the air by means of a screw pressing on the valve of the receiver, a continued stream of air, at the very time of its expansion, passed over the bulb of the thermometer. This experiment was four times repeated in the presence of many observers, and uniformly sunk the thermometer from five to seven degrees. During the time of condensing the air into the receiver, there was a great difference in the heat, as perceived by the hand, at the two ends of the condensing syringe; that next the air-globe was almost painful to the touch; and the globe itself became hotter than could have been expected from its contact with the syringe. Add to this, that in exploding an air-gun, the stream of air always becomes visible, which is owing to the cold then produced precipitating the vapour it contained; and if this stream of air had previously been more condensed, or in greater quantity, so as not instantly to acquire heat from the common atmosphere in its vicinity, it would probably have fallen in snow, as in the fountain of Hiero, mentioned below.

2. About twelve or fourteen years ago, by the assistance of Mr. Warltire, a celebrated itinerant teacher of philosophy, a thermometer was placed in the receiver of an air-pump, and some time being allowed, that it might accurately adapt itself to the heat of the receiver, the air was hastily exhausted;

during which the mercury of the thermometer sunk two or three degrees, and after some minutes regained its previous height. In November 1787, by the assistance of my very ingenious friend Mr. Forester French, the above experiment was repeated; but with this difference, that the thermometer was open at the top; so that the diminution of external pressure could not affect the dimensions of the bulb; and the result was the same, the mercury in the thermometer sunk two or three degrees, and gradually rose again. Does not this shew that the air in the receiver, being expanded during the exhaustion, attracted or absorbed heat from the mercury in the thermometer?

Both during the exhaustion, and during the re-admission of the air into the receiver, a steam was regularly observed to be condensed on the sides of the glass, which in both cases was in a few minutes re-absorbed. This steam must have been precipitated by its being deprived of its heat by the expanded air: if it could have happened from any other cause, the vapour could not, in both situations, viz. of exhaustion, and of re-admission, have been taken up again.

3. In December 1784, with the assistance of Mr. Fox, the following experiment was carefully made. A hole, about the size of a crow-quill, was bored into a large air-vessel, placed at the commencement of the principal pipe in the waterworks which supply the town of Derby. The water from four pumps, which are worked by a water-wheel, is first thrown into the lower part of this air-vessel, and from thence rises to the top of St. Michael's Church into a reservoir, which may be about thirty-five or forty feet above the level of the air-vessel.

Two thermometers were previously suspended on the leaden air-vessel, that they might become of the same temperature with it; and, as soon as the hole was opened, had their bulbs reciprocally applied so as to receive the stream of air; and the mercury in both of them sunk two divisions, or four degrees. This sinking of the mercury in the thermometers could not be ascribed to any evaporation of moisture from their surfaces, because it was seen, both in exhausting and re-admitting the air into the exhausted receiver, that the vapour which it previously contained was deposited during its expansion.

4. There is a very curious phaenomenon observed in the fountain of Hiero, constructed on a very large scale in the

Chemnicensian mines in Hungary, which is very similar to the experiments above related. In this machine the air, in a large vessel, is compressed by a column of water 260 feet high; a stop-cock is then opened, and as the air issues out with great vehemence, and, in consequence of its previous condensation, becomes immediately much expanded, the moisture it contained is not only precipitated, as in the exhausted receiver above-mentioned, but falls down in a shower of snow, with icicles, adhering to the nosel of the cock. This remarkable circumstance is described at large, with a plate of the machine, in the Philosophical Transactions for 1761, Vol. LII.

5. From the four experiments already related; first, of the mercury sinking in the thermometer, by being exposed to the stream of air from an air-gun; secondly, from its sinking in the receiver of an air-pump, during the time of exhausting it; thirdly, from its sinking when exposed to a stream of air from the air-vessel of a water-engine; and, lastly, from the curious phaenomenon of snow and ice being produced by the stream of expanding air from the fountain of Hiero in an Hungarian mine; there is good reason to conclude, that in all circumstances, when air is mechanically expanded, it becomes capable of attracting the fluid matter of heat from other bodies in contact with it.

Coldness of the Summits of Mountains

Now, as the vast region of air which surrounds our globe is perpetually moving along its surface, climbing up the sides of mountains, and descending into the vallies; as it passes along, it must be perpetually varying its degree of heat, according to the elevation of the country it traverses: for in rising to the summits of mountains it becomes expanded, having so much of the pressure of the super-incumbent air taken away, and, when thus expanded, it attracts or absorbs heat from the mountains in contiguity with it; and when it descends into the valley, and is again compressed into less compass, it again gives out the heat is has acquired to the bodies it becomes in contact with.

The same thing must happen in respect to the higher regions of the atmosphere, which are regions of perpetual frost, as was always suspected, and has of late been demonstrated by the aerial navigators. When large districts of air from the lower parts of the atmosphere are raised two or three miles high, they become so much expanded by the great diminution of the pressure over them, and thence become so cold, that hail or snow is produced from the precipitated vapour, if they contain any: and as there is, in these high provinces of the atmosphere, nothing else for the expanded air to acquire heat from, after the precipitation of its vapour, the same degree of cold continues, till the air, on descending to the earth, acquires again its former state of condensation and of warmth.

The Andes, almost under the line, rests its base on burning sands; about its middle height is a most pleasant and temperate climate, covering an extensive plain, on which is built the city of Quito; while its forehead is incircled with eternal snow, co-eval perhaps with the elevation of the mountain: yet, according to the accounts of Ulloa, these three discordant climates seldom intrench much upon each other's territories. The hot winds below, if they ascend, become cooled by their expansion, and hence cannot affect the snow upon the summit; and the cold winds, that sweep the summit, become condensed as they descend, and of temperate warmth, before they reach the fertile plains of Quito. . . .

The Devaporation of aerial Moisture

As heat appears to be the principal cause of evaporation, as well as of solution, and of fluidity in general, the privation of heat may be esteemed the principal cause of devaporation: for though the air may, by its own power of attraction, or by means of the electricity it may contain, dissolve and suspend a portion of water, as water dissolves and suspends a portion of salt; yet, by the application of cold, these are respectively precipitated; and therefore heat may be assumed as the immediate cause of these solutions. Add to this, that water boils *in vacuo* with less heat; that is, it evaporates *in vacuo* faster or easier than in the open air, and therefore the attractive power of the atmosphere does not seem necessary to evaporation.

Now, when the barometer sinks (from whatever cause not yet understood this may happen), the lower stratum of air becomes expanded by its elasticity, being released from a part of the super-incumbent pressure, and, in consequence of its expansion, robs the vapour which it contains of its heat; whence that vapour becomes condensed, and is precipitated in showers, as is visible in the receiver of an air-pump above mentioned.

There are, however, two other curious circumstances belonging to the devaporation of water, which have not been perhaps much attended to.

First, that the deduction of a small quantity of heat from a cloud or province of vapour, compared with the quantity of heat which was necessary to raise that vapour from water, will devaporate the whole. This circumstance is evident in the operation of common steam-engines, in which a small jet of water, whose heat is often above 48 degrees, perpetually devaporates the steam raised by a comparatively very great quantity of heat under the boiler. This difficult problem is explicable from the principles before established: if a small part of a province of vapour be suddenly condensed, a vacuity takes place, and the contiguous walls of vapour expand themselves into this vacuity; and thus a large area of vapour, perhaps of many miles in circumference, becomes more or less expanded; by this expansion cold is produced (that is, its capacity of receiving heat is increased), and the whole is devaporated.

This very circumstance exactly takes place in the famous steam-engine of Mess. Watt and Boulton; which, from the happy combination of chemical and mechanic power, may justly be esteemed the first machine of human invention. In this excellent machine, after the cylinder is filled with steam, a communication is opened between this reservoir of steam and a small cell, which is kept cold by surrounding water, and free from air by an air-syringe adapted to it. What then happens? The corner of the steam in the cylinder next to this vacuum (with which it now communicates) rushes into it, and the whole steam in the cylinder is thus suddenly expanded, and instantly devaporated: whence the very quick reciprocations of the piston; and that, though the cylinder itself is always kept as hot as boiling water, that is, as hot as the steam was previous to its devaporation.

Something very similar to this is often seen at the commencement of thunder-storms; a small black cloud at first appears, in a few minutes the whole heaven is covered with condensing vapour, and the accumulation or escape of electric matter seems to be rather the consequence than the cause of this sudden and general devaporation. . . .

Conclusion

1. When a small portion of air, suppose a few acres, becomes suddenly contracted into a less compass, either by incidental cold, or by any other cause not yet understood (as the combination of dephlogistic and inflammable gases), the air next in vicinity suddenly expands itself to occupy the vacuity; and by its expansion produces cold and devaporates, and then becomes compressible into less space than it occupied before it parted with its vapour. This then gives occasion to the next circumambient portion of air to go through the same process, that is, to expand, attract the heat from its vapours, devaporate, and then become compressible into less space; and thus, from a small and partial contraction or diminution of air, it seems possible to devaporate a great province.

2. The vapour of a great province of air being thus condensed, would leave a great vacuity in this part of the atmosphere, which would be supplied by winds rushing in on all sides. Suppose this to happen to the north of our climate, a south-west wind would be produced here, which is otherwise very difficult to understand: and if it should ever be in the power of human ingenuity to govern the course of the winds, which probably depends on some very small causes; by always keeping the under currents of air from the S.W. and the upper currents from the N.E. I suppose the produce and comfort of this part of the world would be doubled at least to its inhabitants, and the discovery would thence be of greater utility than any that has yet occurred in the annals of mankind.
(*Philosophical Transactions*, 78, 43–52 (1788))

Darwin has a great deal more to say about the atmosphere in the Notes to his poems: his interest extended into the stratosphere and far beyond. He divided the atmosphere into three strata:

There seem to be three concentric strata of our incumbent atmosphere; in which, or between them, are produced four kinds of meteors; lightning, shooting stars, fire-balls, and northern lights. First, the lower region of air . . . which may

> extend from one to three or four miles high. In this region the common lightning is produced. . . .
>
> The second region of the atmosphere I suppose to be that which has too little tenacity to support condensed vapour or clouds; but which yet contains invisible vapour, or water in aerial solution. In this stratum it seems probable that the meteors called shooting stars are produced; and that they consist of electric sparks, or lightning, passing from one region to another of these invisible fields of aero-aqueous solution. . . .
>
> The second region or stratum of air terminates I suppose where the twilight ceases to be refracted, that is, where the air is 3000 times rarer than at the surface of the earth; and where it seems probable that the common air ends; and is surrounded by an atmosphere of inflammable gas tenfold rarer than itself. In this region I believe fire-balls sometimes to pass, and at other times the northern lights to exist. One of these fire-balls . . . was estimated to be between 60 and 70 miles high, and to travel 1000 miles at the rate of about 20 miles in a second. (*Economy of Vegetation*, Notes, pp. 1–2)

Darwin suggests that fireballs are caused by an electrical discharge—rather like ball lightning. In another note (to Canto I, line 126), Darwin states that the air is probably 4,000 to 10,000 times rarer than at sea level at a height of 40 to 50 miles, which implies that it is 3,000 times rarer at a height of about 35 miles.

Darwin's picture of the atmosphere is in several ways more correct than the models favoured by scientists for 150 years afterwards. To begin with, Darwin was right, and for the right reasons, in thinking that hydrogen ('inflammable gas') would be the main component of the outermost atmosphere, or 'exosphere', though Darwin placed the base of the exosphere far too low. However, his mistake is less serious than that of orthodox scientists in the first half of the twentieth century, who thought that all the hydrogen had escaped. We now believe it is generated from water vapour high in the atmosphere under the action of the sun's ultra-violet light. Darwin's second correct suggestion is probably pure luck. Until about 1950 the density of the atmosphere at heights above 30 miles was not well known, and Darwin's assumption that at a height of 35 miles the air is 3,000 times rarer than at sea level was not based on reliable facts: nevertheless his value is correct.

Darwin was also correct in recognizing that the aurora was electrical, and belonged to the outermost layer of the atmosphere, above 35 miles. As usual, his views went unheeded by those who could have benefited, and the auroral measurements in the International Polar Year of 1882 were ruined because the stations were sited on the assumption that the aurora was only 5 miles high.

Darwin never seems to have considered the idea that fireballs and shooting-stars are caused by debris from outer space burning up in the atmosphere. But if this eighteenth-century blind spot is accepted as inevitable, Darwin's explanation, that the meteors were caused by electrical discharges, is perhaps the best that can be devised. Darwin was sometimes accused of being too speculative: yet here, when he deliberately embarks on speculation—

> There has not yet been even a conjecture which can account for these appearances! One I shall therefore hazard, which, if it does not inform, may amuse the reader.

—he was too cautious to hit upon the correct answer, even though he had the right facts available, namely, that the fireballs appear at a height of 60–70 miles and move at 20 miles per second.

Darwin seems to have discovered another feature of the upper atmosphere usually regarded as a twentieth-century prerogative. This is the night airglow, which is strongest in the early part of the night and is caused by the faint flashes of light emitted by atoms in the high atmosphere as they return to normal after having been raised to an excited state by the sunlight during the day. The discovery of the airglow is usually attributed to Yntema (1909): but Darwin correctly notes that

> The light of the evening, at the same distance from noon, is much greater, as I have repeatedly observed, than the light of the morning: this is owing, I suppose, to the phosphorescent quality of almost all bodies, in a greater or less degree, which thus absorb light during the sun-shine, and continue to emit it again for some time afterwards, though not in such quantity as

> to produce apparent scintillations. (Note to *Loves of the Plants*, IV 45)

It must remain questionable whether Darwin really did see the airglow; for, although the airglow often emits more light than all the stars combined, it is very difficult to sense with the eye from the ground. (It has only been seen in its full glory by astronauts.) However, Darwin may possibly have sensed the airglow, for there was much less artificial lighting in his day, and even his detractors, like Mrs. Schimmel-Penninck, had to admit his exceptional power of detecting the smallest deviations from the normal processes of nature.

We are now so familiar with cold and warm fronts on weather charts that many people will probably be surprised that their importance was not generally recognized until the early decades of the twentieth century. But Darwin, helped no doubt by the pointer of the wind vane on the ceiling of his study, noted the existence and importance of these sharp changes in wind direction and temperature. An entry in his Commonplace Book in 1777, near the diagram of the machine for measuring the North–South airflow, reads as follows:

> Certain it is that in change of the wind from N.E. to S.W. the air becomes instantly warm, before it can have moved half over this island: and on the contrary when it changes from S.W. to N.E. it becomes instantly cold before it has blown more than an hour or two, so that the change of the direction of the wind evidently changes its tendency to absorb or to give out heat. (Commonplace Book, p. 45)

Darwin's interest in meteorology continued throughout his life, and one of the Notes to *The Economy of Vegetation* is a 9,000-word essay on Winds, including a meteorological journal. This essay has many fascinating speculations about the world-wide air circulation, in which Darwin shows full awareness of the importance of vertical motions as well as horizontal winds.

Despite his many contributions to the theory and practice of meteorology, Darwin is not among the seventy-four pioneers of meteorology listed in Napier Shaw's standard work *Meteorology in History*. So fickle is fame that Darwin is best known to meteorologists as the author of some weather verses 'Signs of Rain': and the final irony is that these verses were actually written by Jenner.[48]

We must now leave the atmosphere, though Darwin has a great deal more to say about it, and return to Earth, which interested him just as much. The notes on the Earth's interior in *The Economy of Vegetation* run to about 15,000 words and are almost a treatise in themselves. At the end Darwin provides a summary of Earth history, which I quote in shortened form:

1. The earth was projected along with the other primary planets from the sun. . . .

2. The rotation of the earth round its axis was occasioned by its greater friction or adhesion to one side of the cavity from which it was ejected; and from this rotation it acquired its spheroidical form. As it cooled in its ascent from the sun, its nucleus became harder, and its attendant vapours were condensed, forming the ocean.

3. The masses or mountains of granite, porphery, basalt, and stones of similar structure, were a part of the original nucleus of the earth; or consist of volcanic productions since formed.

4. On this nucleus of granite and basaltes, thus covered by the ocean, were formed the calcareous beds of limestone, marble, chalk, spar, from the exuviae of marine animals; with the flints, or chertz, which accompany them. And were stratified by their having been formed at different and very distant periods of time.

5. The whole terraqueous globe was burst by central fires; islands and continents were raised, consisting of granite or lava in some parts, and of limestone in others; and great vallies were sunk, into which the ocean retired.

6. During these central earthquakes the moon was ejected from the earth, causing new tides; and the earth's axis suffered some change in its inclination, and its rotatory motion was retarded.

7. On some parts of these islands and continents of granite or limestone were gradually produced extensive morasses from the recrements of vegetables and of land animals; and from these morasses, heated by fermentation, were produced clay, marle, sandstone, coal, iron (with the bases of variety of acids); all

which were stratified by their having been formed at different, and very distant periods of time. . . .

11. In several parts of the earth's surface subsequent earthquakes, from the fermentation of morasses, have at different periods of time deranged the position of the matters above described. Hence the gravel, which was before in the beds of rivers, has in some places been raised into mountains, along with clay and coal strata which were formed from morasses and washed down from eminences into the beds of rivers or the neighbouring seas, and in part raised again with gravel or marine shells over them. (From *Economy of Vegetation*, Note XXIV)

Several of the topics mentioned, such as the origin of the Earth and Moon, are still in dispute; on others Darwin is utilizing theories variously proposed by Buffon, Whitehurst and Hutton, but he is usually more correct than any of these three. His account provides a foretaste of Lyell's *Principles of Geology* and stands out from among the erroneous theories current at the time. Most of those who made pronouncements on geology or on the variability of species had to square their theories with the preconceived notion of a wise Creator. Darwin was quite free from this religious straitjacket, which allowed only the absurdly short time of 6,000 years for the complete processes of geology and evolution. Consequently Darwin was able to adopt the naturalistic modern approach, and give basically correct interpretations of both geological and biological evolution.

Darwin's footnotes to the *Economy of Vegetation* are longer than the poem, and cover almost every branch of science and technology. To back up these footnotes he has about 50,000 words of additional notes: the subjects include meteors, primary colours, coloured clouds, comets, the sun's rays, the Earth's interior, heat, the momentum of light, luminous insects, phosphorus, the steam engine (Watt provided material for this section), frost, electricity, calcareous earth, morasses, iron, flint, clay, enamels, the Portland Vase (Darwin gave the best explanation of the scenes portrayed on it), coal, granite, evaporation, springs, shellfish, sturgeon, oil on water, ship-worm, maelstrom, glaciers, winds, and finally various notes on vegetation, which provide a foretaste of *Phytologia*.

In *The Loves of the Plants* the notes cover a narrower field, being devoted mainly to the flowers he successively (and often excessively) describes. However, Darwin also offers a few new discoveries or interpretations. For example, there is his account of animal camouflage:

> The colours of insects and many smaller animals contribute to conceal them from the larger ones which prey upon them. Caterpillars which feed on leaves are generally green; and earthworms the colour of the earth which they inhabit; butterflies, which frequent flowers, are coloured like them; small birds which frequent hedges have greenish backs like the leaves, and light-coloured bellies like the sky, and are hence less visible to the hawk who passes under them or over them. Those birds which are much amongst flowers, as the goldfinch (Fringilla carduelis), are furnished with vivid colours. The lark, partridge, hare, are the colour of dry vegetables or earth on which they rest. And frogs vary their colour with the mud of the streams which they frequent; and those which live on trees are green. Fish, which are generally suspended in water, and swallows, which are generally suspended in air, have their backs the colour of the distant ground, and their bellies of the sky. In the colder climates many of these become white during the existence of the snows. (From Note to *Loves of the Plants*, I 375)

This Note was translated into verse by Alfred Noyes in *The Torch-Bearers* (pp. 244–5).

Darwin also emphasizes the occurrence of mimicry in plants:

> There is another plant, Medicago polymorpha, which may be said to assume a great variety of shapes; as the seed-vessels resemble sometimes snail-horns, at other times caterpillars with or without long hair upon them; by which means it is probable they sometimes elude the depredations of those insects. . . .
>
> In the Persian Iris the end of the lower petal is purple, with white edges and orange streaks, creeping, as it were, into the mouth of the flower like an insect; by which deception in its native climate it probably prevents a similar insect from plundering it. . . . There is a curious contrivance attend-

> ing the Ophrys, commonly called the Bee-orchis, and the Fly-orchis, with some kinds of the Delphinium, called Bee-lark-spurs, to preserve their honey; in these the nectary and petals resemble in form and colour the insects which plunder them: and thus it may be supposed they often escape these hourly robbers, by having the appearance of being pre-occupied. .
>
> The cyprepedium from South America . . . has a large globular nectary about the size of a pidgeon's egg of a fleshy colour, and an incision or depression on its upper part, much resembling the body of the large American spider. . . . The similitude of this flower to this great spider seems to be a vegetable contrivance to prevent the humming-bird from plundering its honey.
>
> (*Loves of the Plants*, pp. 177, 181, 26; *Economy of Vegetation*, pp. 201–2)

Though Darwin was wrong about honey, he seems to have been the first to describe mimicry in plants and appreciate its significance.[49]

Darwin also gives some important facts about the dispersal of seeds, in his note on *Cassia*:

> The seeds are black, the stamens gold-colour. This is one of the American fruits which are annually thrown on the coasts of Norway; and are frequently in so recent a state as to vegetate, when properly taken care of. The fruit of the anacardium, cashew-nut; of cucurbita lagenaria, bottlegourd; of the mimosa scandens, cocoons; of the piscidia erythrina, logwood-tree; and cocoa-nuts are . . . amongst these emigrant seeds. . . .
>
> Sir Hans Sloane has given an account of four kinds of seeds, which are frequently thrown by the sea upon the coasts of the islands of the northern parts of Scotland . . . which seeds are natives of the West Indies, and seem to be brought thither by the gulf-stream described below. One of these is called, by Sir H. Sloane, Phaseolus maximus perennis, which is often also thrown on the coast of Kerry in Ireland; another is called, in Jamaica, Horse-eye-bean; and a third is called Niker in Jamaica.
>
> (*Loves of the Plants*, p. 124)

If Charles Darwin had read this, he would perhaps not have written in the *Origin of Species*, seventy years later: 'Until I tried, with Mr. Berkeley's aid, a few experiments, it was not even known how far seeds could resist the injurious action of sea-water.' (Chapter XII)

The additional notes to *The Temple of Nature* are another mine of information and speculation on all branches of science.

In the first note Darwin discusses spontaneous vitality. He believes that microscopic life can in certain circumstances arise from non-living material within a few days. The experiments he relies upon were later discredited, and for many years spontaneous vitality became a dead issue. However it has recently been re-vitalized, because most biologists now accept that life began spontaneously, probably in primeval oceans; so the thesis is respectable and the only question now is '*How long* does it take life to arise spontaneously?' In his note Darwin directs attention to the importance of the microscope, then much neglected, and writes:

> The knowledge of microscopic animals is still in its infancy . . . many more classes, as well as innumerable individuals, may be discovered by improvements of the microscope, as Mr. Herschel has discovered so many thousand stars, which were before invisible, by improvements of the telescope. . . . I hope that microscopic researches may again excite the attention of philosophers, as unforeseen advantages may probably be derived from them, like the discovery of a new world. *Temple of Nature*, Additional Notes, pp. 9, 11)

In a subsequent note on reproduction Darwin writes:

> There is one curious circumstance of animal life analogous in some degree to this wonderful power of reproduction; which is seen in the propagation of some contagious diseases. Thus one grain of variolous matter, inserted by inoculation, shall in about seven days stimulate the system into unnatural action; which in about seven days more produces ten thousand times the quantity of a similar material thrown out on the skin in pustules! (*Temple of Nature*, Additional Notes, p. 38)

Put these last two quotations together and we have all the ingredients of the germ theory of disease. Darwin appreciated the importance of the microscopic 'animalcules' and the great power of successive reproduction. Yet he failed to discover the germ theory of disease, because he was too strongly committed to his theories about the four faculties of the sensorium.

These Additional Notes to *The Temple of Nature* run to over 120 quarto pages. Darwin provides a perceptive review of 'Old Age and Death', but offers no elixirs. In a discussion of Hereditary Diseases, he warns that

> it is often hazardous to marry an heiress, as she is not unfrequently the last of a diseased family. (p. 45)

He offers a two-fluid theory of Electricity and Magnetism, with descriptions of numerous experiments. His 'Analysis of Taste' is a wide-ranging survey that includes the idea of a melody of colours, 'luminous music' as Darwin calls it, the forerunner of kinetic painting. He has long and learned discussions on 'Theory and Structure of Language' and 'Analysis of Articulate Sounds', ending with the description of his speaking-machine already quoted. These longer notes, together with the numerous footnotes to the poem, contain many fascinating items of information about the functioning of nature, but we cannot sample them further.

However, we can hardly end this chapter on Darwin's work in what the eighteenth century called 'natural philosophy' without some mention of his contributions to what the twentieth century calls 'philosophy'. In my view Darwin's main achievement in philosophy is his concept of organic happiness, based on evolutionary progress, which antedates and yet outstrips both Bentham's Utilitarianism and Charles Darwin's evolutionism.

But he also contributed to another and quite different facet of philosophical thought, through his further development of Hartley's materialism. In G. H. Lewes's *History of Philosophy* Darwin receives a whole chapter, as 'one of the psychologists who aimed at establishing the physiological basis of mental phenomena' (Vol. II, p. 356). This theme is taken up by A. Wolf in his standard work *A History of Science, Technology and Philosophy in the Eighteenth Century*, where more space is given to Darwin as a philosopher than as a scientist; and by J. V. Logan, in his book on *The Poetry and Aesthetics of Erasmus Darwin*. Logan provides perhaps the clearest picture of Darwin's attitude:

> He believes that all appetites and all intellectual life can be explained on the basis of physiology. Analyze organic life with its muscles, nerves and fibres, and you will discover the source of not only the physical, but also the intricate moral and emotional life of mankind. . . . It was out of this system of aestho-physiology that Erasmus Darwin's theory of poetry and fine art grew. (p. 21)

In *Zoonomia* Darwin himself states quite emphatically:

> It is certain that all our ideas were originally acquired by our organs of sense; for whatever excites our perception must be external to the organ that perceives it, and we have no other inlets to knowledge but by our perceptions. (*Zoonomia*, I 36)

Darwin's ideas on aesthetics are debatable, and I shall not pursue them further.[50]

13 Summary Justice

'Give your evidence', said the King; 'and don't be nervous, or I'll have you executed on the spot.'

LEWIS CARROLL; *Alice in Wonderland*

UNFORTUNATELY the evidence of Darwin's achievements is too widely dispersed to summarize justly. It is easy to say, without being nervous of contradiction, that he was the finest physician of his time, notable for his kindness to poor patients; a prolific inventor; an intensely sociable man, who would found a Society whenever he had the chance; a much admired poet who versified science with unsurpassed skill, and strongly influenced Wordsworth, Coleridge and Shelley; and a scientist who made advances in subjects ranging between evolution and meteorology. But such generalities die away in the reader's mind like spreading ripples on a pond.

To escape that threatened 'execution on the spot', we need instead an array of concrete examples to place like dragon's teeth right in the path of judge, jury and executioner. So perhaps the best plan is to set aside all generalities: to ignore Darwin's devoted medical work and philanthropy, which absorbed so much of his energy; to ignore also the first half of Coleridge's dictum about Darwin being 'the first *literary* character in Europe', and to concentrate instead on the second half—that he was 'the most original-minded man'.[51] As we have seen in this book, Darwin's originality extended to many different subjects; the list on this and the next page brings together seventy-five subjects in which he was a pioneer, i.e. advanced new and correct ideas, developed new inventions or techniques, etc. The topics are arranged alphabetically, to bring out their full diversity.

1. abolition of slavery
2. adiabatic expansion
3. aesthetics
4. afforestation
5. air travel
6. animal camouflage
7. artesian wells
8. artificial insemination

9. aurorae
10. biological adaptation
11. biological pest control
12. canal lifts
13. carriage design
14. cemeteries
15. centrifugation
16. clouds
17. compressed-air actuators
18. copying-machines
19. educational reform
20. electrical machines
21. electrotherapy
22. evolutionary theory
23. exercise for children
24. fertilizers
25. formation of coal
26. geological stratification
27. hereditary diseases
28. individuality of buds
29. insecticides
30. language
31. light verse
32. limestone deposits
33. manures
34. materialism
35. mental illness
36. microscopy
37. mimicry
38. moon's origin
39. nerve impulses
40. night airglow
41. nitrogen cycle
42. ocular spectra
43. organic happiness
44. origin of life
45. outer atmosphere
46. phosphorus
47. photosynthesis
48. Portland Vase
49. rocket motors
50. rotary pumps
51. secular morality
52. seed-drills
53. sewage farms
54. sexual reproduction
55. speaking-machines
56. squinting
57. steam carriages
58. steam turbines
59. struggle for existence
60. submarines
61. survival of the fittest
62. telescopes
63. temperance
64. timber growing
65. travel of seeds
66. treatment of dropsy
67. ventilation
68. versifying science
69. warm and cold fronts
70. water closets
71. water machines
72. wind-gauges
73. windmills
74. winds
75. women's emancipation

With this concise array of facts on full display, the judge can justly say, 'Ladies and gentlemen of the jury, please consider your verdict. Can you think of any other individual in the last two or three hundred years with seventy-five achievements so notable and so varied? You must be sure that your answer is "yes" before you think of condemning Darwin or his editor to summary execution'.

14 References

NOTE: The title of a book may be shortened and its date omitted, if it appears in the book list which follows these references.

1. A. Seward, *Life of Dr Darwin*, pp. 1–2.
2. R. L. and M. Edgeworth, *Memoirs*, p. 102.
3. See H. W. Dickinson, *Matthew Boulton*.
4. See E. Meteyard, *Life of Josiah Wedgwood*.
5. See D. Clarke, *The Ingenious Mr Edgeworth*.
6. See L. T. C. Rolt, *James Watt*; H. W. Dickinson, *James Watt* (Cambridge Univ. Press, 1936); J. G. Crowther *Scientists of the Industrial Revolution*, pp. 93–172.
7. See J. K. Moilliet, *James Keir*.
8. J. L. Moilliet, *Chemistry and Industry*, No. 10, p. 405 (1966).
9. See R. E. Schofield, *Lunar Society*, and E. Robinson, *The Lunatics*.
10. See D. King-Hele, *Nature*, 212, 229–33 (1966).
11. A. Seward, *Life of Dr Darwin*, pp. 12–14.
12. See S. H. Scott, *The Exemplary Mr Day* (Faber, 1935).
13. See T. W. Peck and K. D. Wilkinson, *William Withering*.
14. See F. W. Gibbs, *Joseph Priestley*.
15. A. Seward, *Life of Dr Darwin*, p. 5.
16. See J. E. Stock, *Life of Thomas Beddoes*.
17. S. T. Coleridge, *Letters*, I 177.
18. See J. Muirhead, *James Watt*, I, clx-clxi, and F. W. Gibbs, *Joseph Priestley*, pp. 154–5.
19. See D. Gardiner, *English Girlhood at School* (O.U.P., 1929), pp. 347–56; B. Simon, *Studies in the History of Education*, pp. 50–6; J. Kamm, *Hope Deferred* (Methuen, 1965) passim; and *Encyclopaedia Britannica*, Vol. 23, article on 'Women, Education of'.
20. J. Wedgwood, letter to Bentley, 8 Nov. 1779 (*Selected Letters*, p. 244).
21. C. Darwin, *Life of Erasmus Darwin*, p. 102.
22. See C. Darwin, *Life of Erasmus Darwin*, p. 84; R. E.

Schofield, *Lunar Society*, pp. 270–4, 435; and J. M. Drachman, *Studies in the Literature of Natural Science* (Macmillan, New York, 1930), p. 161.

23. J. Wedgwood, *Selected Letters*, pp. 243–4.
24. W. J. White, *A History of the Centrifuge in Aerospace Medicine* (Douglas Aircraft Co., Santa Monica, 1964), Chap. 1.
25. See D. King-Hele, *Erasmus Darwin*, pp. 63–96.
26. See C. D. Darlington, *Darwin's place in history*, Chaps. 3 and 11; and D. King-Hele, *Erasmus Darwin*, pp. 79–90.
27. See *Monthly Magazine*, 13, 460 (1802), and A. and N. Clow, *The Chemical Revolution*, p. 490.
28. E. J. Russell, *History of Agricultural Science in Great Britain* (1966), pp. 72–3.
29. See W. A. Riley, *Science*, 73, 475 (1931) and *Encyclopaedia Britannica*, Vol 12, article on 'Insect'.
30. See *Missiles and Rockets*, 25 April 1966, pp. 24–5.
31. S. P. Thomson, *Journ. Soc. Telegraph Eng.*, 17, 576 (1888). See also R. E. Schofield, *Lunar Society*, p. 166.
32. For further information on Darwin's carriages, see M. A. Hopkins, *Dr Johnson's Lichfield*, pp. 92–7; Julia Wedgwood, *The Personal Life of Josiah Wedgwood* (Macmillan, 1915), p. 261; R. L. and M. Edgeworth, *Memoirs*, pp. 100–106.
33. For further details of canal lifts, see C. R. Weaver, *Discovery*, July 1965, p. 57, and C. Hadfield, *Canals of the West Midlands* (David and Charles, 1966).
34. For further discussion of Darwin's minor poems, see J. V. Logan, *The Poetry and Aesthetics of Erasmus Darwin*, pp. 93–109, and Anna Seward, *Life of Dr Darwin*, passim.
35. J. V. Logan, *Poetry and Aesthetics of Erasmus Darwin*, p. 111.
36. See E. Robinson, *Annals of Science*, 10, 314–20 (1954); N. Garfinkle, *Journ. Hist. Ideas*, 16, 376–88 (1955); B. Blackstone, *The Consecrated Urn* (1959); E. Sewell, *The Orphic Voice* (1960); H. W. Piper, *The Active Universe* (1962); D. King-Hele, *Erasmus Darwin*, Chaps. VI–VIII (1963); I. Primer, *Journ. Hist. Ideas*, 25, 58–76 (1964); W. P. Jones, *The Rhetoric of Science* (1966).

37. For further details of Darwin's influence on Wordsworth, see E. Legouis, *The Early Life of William Wordsworth* (1921), pp. 137–8; H. W. Garrod, *Wordsworth* (2nd ed., 1927), pp. 55–6; A. Beatty, *William Wordsworth* (2nd ed., 1927), pp. 57–65; D. King-Hele, *Erasmus Darwin*, pp. 137–40.
38. *The Notebooks of S. T. Coleridge*, ed. K. Coburn (1957), Vol. I, Notes, No. 132.
39. For further details of Darwin's influence on Coleridge, see J. L. Lowes, *The Road to Xanadu*; *The Notebooks of S. T. Coleridge*, ed. K. Coburn (1957); H. W. Piper, *The Active Universe*; D. King-Hele, *Erasmus Darwin*, pp. 140–2. Most of the parallels I have quoted (and others) are to be found in one or other of these books.
40. For further discussion of Darwin's influence on Keats, see B. Blackstone, *The Consecrated Urn.*
41. See *Keats-Shelley Memorial Bulletin*, *XVI*, 26–8 (1965)
42. For further discussion of Darwin's influence on Shelley, see C. Grabo, *A Newton Among Poets*; D. King-Hele, *Shelley: his Thought and Work* and *Erasmus Darwin.*
43. See B. Blackstone, *English Blake* (1949) and *The Consecrated Urn*; E. Robinson, *Annals of Science*, **10**, 314 (1954); D. King-Hele, *Erasmus Darwin*, p. 143; L. T. C. Rolt, *Thomas Telford* (1958); J. E. Stock, *Life of Thomas Beddoes.*
44. See W. E. K. Middleton, *Theories of Rain*, pp. 112–15.
45. For an account of the foxglove controversy, very much from Withering's side, see R. E. Schofield, *Lunar Society*, pp. 163–5, 307–21.
46. For further discussion of Darwin's work on adiabatic expansion, see R. E. Schofield, *Lunar Society*, pp. 267–9, where there is a photograph of Darwin's apparatus, and W. E. K. Middleton, *Theories of Rain*, pp. 103–5. Darwin's achievement was recognized by later authorities—J. Dalton, *Meteorological Observations and Essays* (2nd ed., Manchester, 1834), pp. 125–32; T. Young, *Course of Lectures*, 2nd ed., Vol. i, p. 485; J. P. Joule, *Philosophical Magazine*, **26**, 369 (1845).
47. For further discussion of Darwin's work on how clouds form, see W. E. K. Middleton, *Theories of Rain*, pp. 105–6.

48. See J. Baron, *Life of Jenner*, I 22–4.
49. See P. C. Ritterbush, *Overtures to Biology*, p. 168.
50. For further discussion of Darwin's philosophical ideas, see J. V. Logan, *Poetry and Aesthetics of Erasmus Darwin*, pp. 21–92.
51. S. T. Coleridge, *Letters*, I 305.

15 Book List

THIS SECTION is in two parts. Part A is a list of Darwin's main published writings. Part B gives the books which, in my opinion, provide the most useful background reading for anyone wishing to learn more about Darwin, his work and his circle of friends.

A. ERASMUS DARWIN'S MAIN PUBLISHED WRITINGS

1. 'The Death of Prince Frederick' [a poem]. 1751: republished in the *European Magazine*, Feb. 1795.
2. 'Remarks on the Opinion of Henry Eeles, Esq., concerning the Ascent of Vapour'. *Philosophical Transactions of the Royal Society*, 50, 240–54 (1757).
3. 'An uncommon Case of an Haemoptysis'. *Phil. Trans. Roy. Soc.*, 51, 526–9 (1760).
4. 'Experiments on Animal Fluids in the exhausted Receiver'. *Phil. Trans. Roy. Soc.*, 64, 344–349 (1774).
5. 'A New Case in Squinting'. *Phil. Trans. Roy. Soc.*, 68, 86–96 (1778).
6. 'An Account of an artificial Spring of Water'. *Phil. Trans. Roy. Soc.*, 75, 1–7 (1785).
7. 'An Account of the successful use of Foxglove in some Dropsies, and in the pulmonary Consumption'. *Medical Transactions*, 3, 255–86. (1785).
8. 'Frigorific Experiments on the mechanical Expansion of Air'. *Phil. Trans. Roy. Soc.*, 78, 43–52 (1788).
9. 'Of the Medicinal Waters of Buxton and Matlock', pp. 256–75 in J. Pilkington's *A View of the Present State of Derbyshire* (London, 1789).
10. *The Botanic Garden*

    Part I, *The Economy of Vegetation* (London, J. Johnson, 1791)

    Part II, *The Loves of the Plants* (London, J. Johnson, 1789).

    Later English editions: *Part I.* 2nd edition, 1791. 3rd, 1795. 4th, 1799. 5th, 1806–7. 6th, 1824. *Part II.*

2nd, 1790. 3rd, 1791. 4th, 1795. 5th, 1799. 6th, 1806–7. 7th, 1824. Also Irish editions (1793–6).
American editions: New York, 1798. 2nd ed., 1807.
Translations. French: Paris, 1800. Italian: Milan, 1805. Portuguese: Lisbon, 1803–4.
Beauties of the Botanic Garden [selections]. London, Cadell, 1805. New York, Longworth, 1805.

11. *Zoonomia; or, The Laws of Organic Life*
Part I: London, J. Johnson, 1794.
Parts I–III: London, J. Johnson, 1796 (2nd ed., corrected), 2 vols.
Third edition (with further corrections), London, J. Johnson, 1801. (4 vols.).
Irish edition: Dublin, P. Byrne, 1800.
American editions: New York, 1796. Philadelphia, 1797 and 1818. Boston, 1803 and 1809.
Translations. German: Hanover, 1795–7. Italian: Venice, 1803–6. Also French edition.

12. *A Plan for the Conduct of Female Education in Boarding Schools*
Derby, for J. Johnson, 1797. Other editions: Dublin, J. Chambers, 1798; Philadelphia, J. Ormond, 1798.

13. *Phytologia; or, The Philosophy of Agriculture and Gardening*
London, J. Johnson, 1800. Other editions: Dublin, P. Byrne, 1800; and a German edition.

14. *The Temple of Nature; or, The Origin of Society*
London, J. Johnson, 1803. 2nd ed., London, J. Johnson, 1806–7. 3rd ed., London, J. Jones, 1825. American editions: New York, 1804; Baltimore, 1804. *Russian translation:* 1911, in Journal of Ministry of National Education. In book form, Moscow, 1954. 2nd ed., Moscow, 1960.

A fuller bibliography of the poems can be found in J. V. Logan, *The Poetry and Aesthetics of Erasmus Darwin*, pp. 149–53. Some additional scientific publications, including three books of which Darwin was editor, can be found in the list given by R. E. Schofield, *The Lunar Society*, pp. 442–4. For details of the later editions of *Botanic Garden* and *Zoonomia* I have relied mainly on information given by Logan or Schofield.

B. FURTHER READING

(The place of publication is London, unless otherwise stated.)

Nora BARLOW, 'Erasmus Darwin'. Pages 85–98 in *Notes and Records of the Royal Society of London*, vol. 14, No. 1, 1959.

J. BARON, *The Life of Edward Jenner*. 2 vols. Colburn, 1838.

B. BLACKSTONE, *The Consecrated Urn*. Longmans, 1959.

Eileen BLACKWELL, 'The Life, Times and Work of Erasmus Darwin'. Unpublished thesis (1949). Available at London University Library.

T. BROWN, *Observations on the Zoonomia of Erasmus Darwin*. Edinburgh, 1798.

D. CLARKE, *The Ingenious Mr Edgeworth*. Oldbourne, 1965.

Archie and Nan CLOW, *The Chemical Revolution*. Batchworth, 1952.

S. T. COLERIDGE, *Collected Letters* (editor E. L. Griggs). Oxford Univ. Press. Vols. I and II, 1956. Vols. III and IV, 1959.

J. G. CROWTHER, *Scientists of the Industrial Revolution*. Cresset Press, 1962.

C. D. DARLINGTON, *Darwin's Place in History*. Blackwell, Oxford, 1959.

Charles DARWIN, *The Life of Erasmus Darwin*, together with an Essay on his scientific works by Ernst Krause. Murray, 1879. (2nd ed., 1887).

Charles DARWIN, *Autobiography*. Edited, and with Appendices, by Nora Barlow. Collins, 1958.

H. W. DICKINSON, *Matthew Boulton*. Univ. Press, Cambridge, 1936.

R. L. and Maria EDGEWORTH, *Memoirs of Richard Lovell Edgeworth*, 3rd ed., Bentley, 1844.

F. W. GIBBS, *Joseph Priestley*. Nelson, 1965.

C. GRABO, *A Newton among Poets*. Univ. of N. Carolina Press, Chapel Hill, N.C., 1930.

Christina C. HANKIN (editor), *Life of Mary Anne Schimmel-Penninck*. 2 vols. Longmans, 1858.

Mary A. HOPKINS, *Dr Johnson's Lichfield*. Owen, 1956.

D. KING-HELE, *Shelley: his Thought and Work*. Macmillan, 1960.

D. King-Hele, *Erasmus Darwin*. Macmillan, 1963.

G. H. Lewes, *The History of Philosophy*. 3rd ed., 2 vols. Longmans, 1867.

J. V. Logan, *The Poetry and Aesthetics of Erasmus Darwin*. Princeton Univ. Press, Princeton, 1936.

J. L. Lowes, *The Road to Xanadu* (1927). Vintage Books, New York, 1959.

Eliza Meteyard, *The Life of Josiah Wedgwood*. 2 vols. Hurst & Blackett, 1865–6.

Eliza Meteyard, *A Group of Englishmen*. Longmans, 1871.

W. E. K. Middleton, *A History of the Theories of Rain*. Oldbourne, 1965.

J. K. Moilliet (editor), *Life and Correspondence of James Keir*. Privately printed, 1868.

J. P. Muirhead, *The Origin and Progress of the Mechanical Inventions of James Watt*. 3 vols. Murray, 1854.

A. Noyes, *The Torch-Bearers*. Sheed & Ward, 1937.

Hesketh Pearson, *Doctor Darwin*. Dent, 1930 (Penguin, 1943).

Hesketh Pearson, *Extraordinary People*. Heinemann, 1965.

Karl Pearson, *The Life, Letters and Labours of Francis Galton*. 4 vols. Univ. Press, Cambridge, 1914–30.

T. W. Peck and K. D. Wilkinson, *William Withering of Birmingham*. Wright, Bristol, 1950.

H. W. Piper, *The Active Universe*. Athlone Press, 1962.

I. Primer, 'Erasmus Darwin's *Temple of Nature*'. *Journal of the History of Ideas*, 25, 58–76, 1964.

L. Rice-Oxley (editor), *Poetry of the Anti-Jacobin*. Blackwell, Oxford, 1924.

P. C. Ritterbush, *Overtures to Biology*. Yale Univ. Press, New Haven, 1964.

L. T. C. Rolt, *James Watt*. Batsford, 1962.

R. E. Schofield, *The Lunar Society of Birmingham*. Oxford Univ. Press, 1963.

Anna Seward, *Memoirs of the Life of Dr Darwin*. Johnson, 1804.

Elizabeth Sewell, *The Orphic Voice*. Yale Univ. Press, New Haven, 1960.

B. Simon, *Studies in the History of Education*. Lawrence & Wishart, 1960.

Barbara M. D. SMITH and J. L. MOILLIET, 'James Keir of the Lunar Society'. Pages 144–54 in *Notes and Records of the Royal Society of London*, vol. 22, 1967.

J. E. STOCK, *Memoirs of the Life of Thomas Beddoes*. Murray, 1811.

Josiah WEDGWOOD, *Selected Letters* (editors A. Finer and G. Savage). Cory, Adams & Mackay, 1965.

H. WILLIAMS, *Great Biologists*. Bell, 1961.

A. WOLF, *A History of Science, Technology and Philosophy in the Eighteenth Century* (1938). 2nd ed., Allen & Unwin, 1952.

Acknowledgements

I SHOULD like to express my sincere thanks to the many people who so kindly helped me with this book. My wife Marie read the manuscript and suggested many important improvements. Professor Hedley Atkins, President of the Royal College of Surgeons, gave me permission to reproduce photographs of items belonging to the Darwin Museum at Down House, namely the portrait of Mrs Pole (Fig. 2), and extracts from Darwin's manuscript Commonplace Book (Figs. 9–14 and 16, and quotations in the text). Dr John Moilliet allowed me to borrow and quote from the unpublished memoirs of Darwin's granddaughter Elizabeth Anne Wheler, which he is now editing with a view to publication, and also provided the silhouette reproduced in Fig. 4. Professor F. G. Young, Master of Darwin College, Cambridge, supplied the portrait of Erasmus Darwin in Fig. 1, and Mr George Darwin gave permission for it to be reproduced. Mr R. H. Greening, who lives at Darwin's house in Lichfield, helped me to obtain a photograph of the house (Fig. 5). Mr John Lowe, Director of the Birmingham City Museum and Art Gallery, allowed me to study Darwin's Commonplace Book at a time when it was on display in the Lunar Society Exhibition. The Manager of the Mint, Birmingham, provided the photograph of the Lunar Society medallion (Fig. 3). Miss Q. Garlick typed a difficult manuscript extremely well. Mr I. Kaye and Mr N. H. Robinson, Librarians at the Royal Society, gave me useful advice and assistance. And last, but certainly not least, Mr Richard Parker and the staff of the Farnham Branch of the Surrey County Library obtained a large number of often obscure books and enabled me to keep them on extended loan. To all these, and to several others who have given advice, I am most grateful.

D.K.

February 1967

Index

Erasmus Darwin's writings are indexed under their titles. Books and poems by other authors are indexed under the author's name. Figures in **bold type** indicate the leading references on a subject.